Graphene Production and Application

Edited by Sadia Ameen,
M. Shaheer Akhtar and Hyung-Shik Shin

Published in London, United Kingdom

IntechOpen

Supporting open minds since 2005

Graphene Production and Application
http://dx.doi.org/10.5772/intechopen.83309
Edited by Sadia Ameen, M. Shaheer Akhtar and Hyung-Shik Shin

Contributors
Sadia Ameen, M. Shaheer Akhtar, Hyung-Shik Shin, Rishikesh Godbole, Alexander S. Dmitriev, Arnaldo César Pereira, Thiago Gabry Barbosa, Ana Elisa Ferreira Oliveira, Ernesto Hernández-Hernández, Perla J. Hernández-Belmares, Monica A. Ceniceros-Reyes, Oliverio S. Rodríguez-Fernández, Pablo Gonzalez-Morones, Ramesh Rudrapati, Nurhafizah Md Disa

First published in London, United Kingdom, 2020 by IntechOpen
IntechOpen is the global imprint of INTECHOPEN LIMITED, registered in England and Wales, registration number: 11086078, 7th floor, 10 Lower Thames Street, London, EC3R 6AF, United Kingdom
Printed in Croatia

British Library Cataloguing-in-Publication Data
A catalogue record for this book is available from the British Library

Additional hard and PDF copies can be obtained from orders@intechopen.com

Graphene Production and Application
Edited by Sadia Ameen, M. Shaheer Akhtar and Hyung-Shik Shin
p. cm.
Print ISBN 978-1-83880-191-5
Online ISBN 978-1-83880-192-2
eBook (PDF) ISBN 978-1-83880-252-3

For EU product safety concerns:
IN TECH d.o.o., Prolaz Marije Krucifikse Kozulić 3, 51000 Rijeka, Croatia, info@intechopen.com or visit our website at intechopen.com.

Meet the editors

Professor Sadia Ameen obtained a PhD in Chemistry in 2008 from Jamia Millia Islamia, New Delhi, India, and then moved to Jeonbuk National University, Republic of Korea. Presently, she is Assistant Professor at Department of Bio-Convergence Science, Jeongeup Campus, Jeonbuk National University. Her current research focuses on dye-sensitized solar cells, perovskite solar cells, organic solar cells, sensors, catalyst, and optoelectronic devices. She is specialized in manufacturing advanced energy materials and nano-composites. She has a Gold Medal in Academics, a merit scholarship, and a Best Researcher Award. She has more than 120 peer-reviewed papers in the field of solar cells, catalyst, and sensors, and has contributed to book chapters, edited books, and is inventor/co-inventor of patents.

Professor M. Shaheer Akhtar completed his PhD in Chemical Engineering, 2008, from Jeonbuk National University, Republic of Korea. Presently, he is Associate Professor at Jeonbuk National University, Republic of Korea. His research interests include photo-electrochemical characterizations of thin film semiconductor nanomaterials, composite materials, polymer-based solid-state films, solid polymer electrolytes and electrode materials for dye-sensitized solar cells (DSSCs), hybrid organic-inorganic solar cells, small molecules-based organic solar cells, and photocatalytic reactions.

Professor Hyung-Shik Shin received Ph.D. in the kinetics of initial oxidation Al (111) surface from the Cornell University, USA, in 1984. He is Professor in School of Chemical Engineering, Jeonbuk National University and also the President of Korea Basic Science Institute (KBSI), Gwahak-ro, Yuseong-gu, Daejon, Republic of Korea. He has been promising researcher and visited several universities as visiting professor/invited speaker worldwide. He is an active executive member of various renowned scientific committees such as KiChE, copyright protection, KAERI, etc. He has extensive experience in electrochemistry, renewable energy sources, solar cells, organic solar cells, charge transport properties of organic semiconductors, inorganic-organic solar cells, biosensors, chemical sensors, nano-patterning of thin film materials, and photocatalytic degradation.

Contents

Preface

Graphene is a very thin material with thickness as little as a single layer of carbon atoms positioned in a honeycomb-shaped lattice. It is the architectural basis of several other allotropes of carbon. The incredibility of graphene lies in its inherent potential to revolutionize modern technology through a multitude of astonishing mechanical, electronic, and optical peculiarities. Since the 2004 discovery that graphene can be isolated in a single layer from graphite, it has stimulated a whirlwind of research worldwide. Nowadays, graphene is beginning to play a central role in the advancement of modern technology. It can be used in a wide range of advanced technologies such as batteries, solar cells, supercapacitors, light-emitting diodes, displays, photocatalysis, hydrogen storage, sensors, tribology, thermal management of devices, and so on. Graphene has also demonstrated the ability to synergize other host materials to enhance their performance when used in combination. Based on the all-inclusive applicability of this remarkable material, graphene has attracted tremendous attention from scientists and engineers to invent and develop novel methods to grow it in the laboratory other than mechanical exfoliation from a pile of graphite.

The authors of this book have brought forward a comprehensive collection about the ideas connected to graphene. The books highlights the importance, functionality, and applicability of graphene, particularly in the field of technology, and describes recent developments in its synthesis, characterization, and application. We hope this book will enhance understanding of this exotic material and help those in the field accelerate graphene's use in the advancement of modern technology. We would like to thank all the contributing authors for their knowledge, efforts, and time. We are also grateful to IntechOpen for their kind help in publishing and making this book a reality.

Sadia Ameen
Advanced Materials and Devices Laboratory,
Department of Bio-Convergence Science,
Jeongeup Campus,
Jeonbuk National University,
Republic of Korea

M. Shaheer Akhtar
New & Renewable Energy Material Development Center (NewREC),
Jeonbuk National University,
Republic of Korea

Hyung-Shik Shin
School of Chemical Engineering,
Jeonbuk National University,
Republic of Korea

Korea Basic Science Institute (KBSI),
Daejon, Republic of Korea

Chapter 1

An Introduction to Graphene Materials

Sadia Ameen, Rishikesh Godbole, M. Shaheer Akhtar and Hyung-Shik Shin

Abstract

Graphene, an allotrope of carbon, is the thinnest compound known to human which is a single layer (monolayer) of carbon atoms, tightly bound in a hexagonal honeycomb lattice. Nanosize graphene is known to possess large surface area and shows promising properties in terms of mechanical, electrical, chemical, and magnetism. Graphene and its derivatives are an exciting replacement for the existing nanomaterials, and so, graphene is discovered to be useful in the application of energy conversion and storage, sensing, electronics, photonics, and biomedicine. In this introductory chapter, the potential implementation of graphene and its nanocomposites, along with the characterization techniques employed for graphene, is briefly discussed. We hope this review can inspire more innovative insights into this emerging topic in energy materials.

Keywords: graphene, nanocomposites, devices, electrochemical properties, synthesis

1. Introduction

In recent years, a new material called graphene has emerged in the regime of advanced nanomaterials and is being extensively explored by the scientists and engineers. Graphene is a two-dimensional (2D) allotrope of carbon, wherein sp^2-hybridized carbon atoms are covalently bonded in a hexagonal lattice to form a single sheet of atoms [1]. It has been realized that graphene happens to be the basic building block of other carbon allotropes such as graphite, single-walled/multi-walled carbon nanotubes, and fullerenes. Monolayer graphene is a perfect 2D material owing to its thickness ranging to a single sheet of carbon atoms. It could exist in bilayer or few layer configurations as well. However, as the numbers of layers increases, it transforms to graphite. Graphene has gained immense popularity in the scientific field due to its fascinating properties such as metal-like conductivity, superior chemical stability, optimum blend of high mechanical strength and flexibility, excellent transport properties with very high charge mobility (200,000 $cm^2 V^{-1} s^{-1}$), zero energy band gap, low optical absorption, and a high specific surface area (2630 $m^2 g^{-1}$) [2–4]. These outstanding characteristics make graphene increasingly demanding in a plethora of applications such as photocatalysis, supercapacitors, batteries, solar devices, light emitting diodes, hydrogen storage, sensors, and so on [5–9]. Graphene could also be incorporated in ceramics, metal oxide semiconductors, or conducting polymers to enhance the performance of the host materials by the synergistic effects. These applications offer exciting new opportunities for scientists to devise new strategies and methods for the synthesis and applications of graphene.

The synthesis of a high-quality graphene to be suitable for any kind of application has always been very critical so far, as there is a stringent limitation on its thickness keeping its adhesion, transparency, and continuity unhampered on larger areas. As an example, graphene could be synthesized by exfoliation of graphite flakes (such as HOPG or Kish graphite) or through the annealing of polar SiC surfaces at high temperature of 1200°C. However, these methods face the major limitations of nontransparency and lesser amount of graphene deposition on the substrate [10]. Hence, in recent years, numerous methods have been designed and utilized for efficient synthesis of graphene without letting its special properties getting deteriorated. The chemical vapor deposition (CVD) technique is one of the techniques being successful in synthesizing graphene at 1000°C from hydrocarbon gases like CH_4 and has become one of the popular methods to prepare transparent and uniform graphene thin film on the metal substrates. It is an inexpensive technique for the high-throughput growth of graphene on larger areas. However, it needs one extra step to transfer graphene film to the arbitrary substrates which needs more efforts and becomes tedious. Another variant of this method is known as hot filament chemical vapor deposition (HFCVD), wherein the precursor gases are heated to a higher temperature of around 1500–2000°C with the help of filaments made up of refractory materials such as tungsten. Due to a higher temperature, it is possible to achieve the complete decomposition of precursor gases and initiate the pyrolytic reactions to attain a complete phase formation of the material to be synthesized on the substrate of choice. It is an emerging technique to synthesize high-quality graphene with only a few reports in the literature [11]. However, graphene-like carbon (GLC) thin films could be successfully synthesized using this technique. Herein, the decomposition reaction is initiated among precursor gases like hydrocarbon (CH_4) gas, hydrogen (H_2), and argon (Ar) under the vacuum conditions by flowing them over hot filaments. The carbon species thus condenses onto the desired substrates, kept close to the filaments, and requires phase if formed [12], as shown in **Figure 1**. Further the GLC thin film could be either directly used or transferred onto different substrates for its utilization in a particular application. As presented in **Figure 2**, these films, when tested as a counter electrode in dye-sensitized solar cells (DSSC), showed a relatively high conversion efficiency of ~4.3% [12] and ~6.94% [13]. It is also presented in **Figure 3** that the GLC-Ni nanocomposite films synthesized by the combination of HFCVD and DC sputtering presented a moderate conversion efficiency of ~3.1% [14].

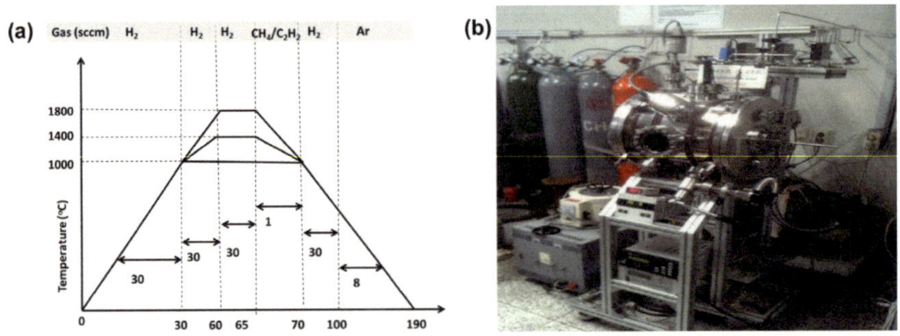

Figure 1.
(a) Schematic illustration showing the parameters for the growth process of graphene-like thin film on FTO glass substrate by HFCVD and (b) photograph of HFCVD system [12].

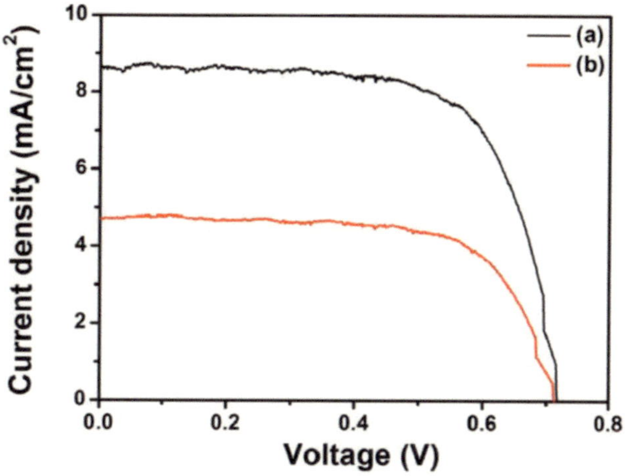

Figure 2.
J-V curves of the DSSCs fabricated with graphene-like thin film counter electrodes obtained from (a) CH_4 and (b) C_2H_2 precursor gases [12].

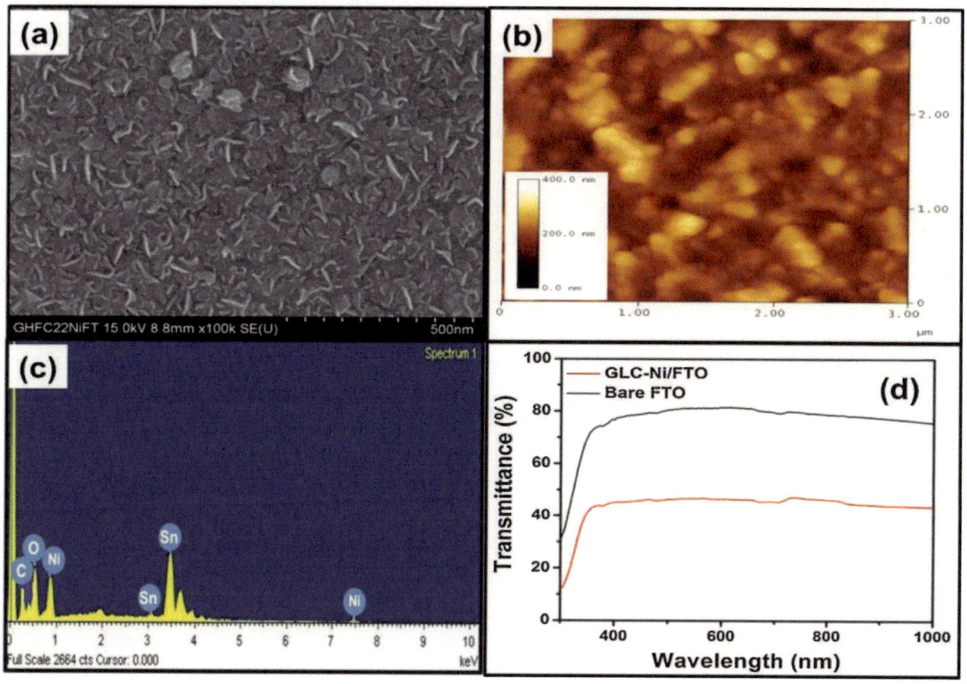

Figure 3.
(a) FESEM, (b) AFM, (c) EDS, and (d) UV-vis spectra at transmittance mode of GLC-Ni/FTO thin film [14].

Graphene could be synthesized by the chemical-based method known as Hummers and Offeman method, in which the standard graphite powder is mixed with sulfuric acid at ~0°C and then potassium permanganate is added to the reaction suspension which raises the temperature to ~20°C. Thereafter, deionized water is added with vigorous stirring which further raises the temperature to ~98°C with subsequent addition of hydrogen peroxide. At first graphene oxide (GO) is obtained after rinsing and drying the precipitates. After reducing this product using hydrazine monohydrate with subsequent washing and drying, graphene oxide gets transformed to graphene. Further, the as-synthesized graphene could be used to make a nanocomposite with polyaniline by a method known as chemical oxidative polymerization of aniline monomer. The composite thus formed could be used in the photocatalytic degradation of rose bengal dye which could serve as an environmental remedy against hazardous industrial effluents [15]. The electrochemical synthesis of as-prepared graphene oxide with aniline monomer was performed and applied as electrode material for the detection of hydrazine chemical. The high sensitivity of ~32.54 × 10^{-5} Acm^{-2} mM^{-1} with response time of 10 s was achieved by Ameen et al. [16]. The graphene oxide was also used for the formation of nanohybrid by using metal oxide such as zinc oxide (ZnO) through the hydrothermal process, carried out at ~90°C for 12 h. Uniform distribution and mixing of ZnO particles and graphene sheets were seen, as shown in **Figure 4**. The synthesized ZnO-GO nanohybrid was utilized as an efficient photocatalyst for the 95% degradation of crystal violet (Cv)-dye in 80 min, as displayed in **Figure 5**. The fast Cv dye degradation

Figure 4.
FESEM images of (a) ZnO nanoparticles (b) ZnO-GO nanohybrid [17].

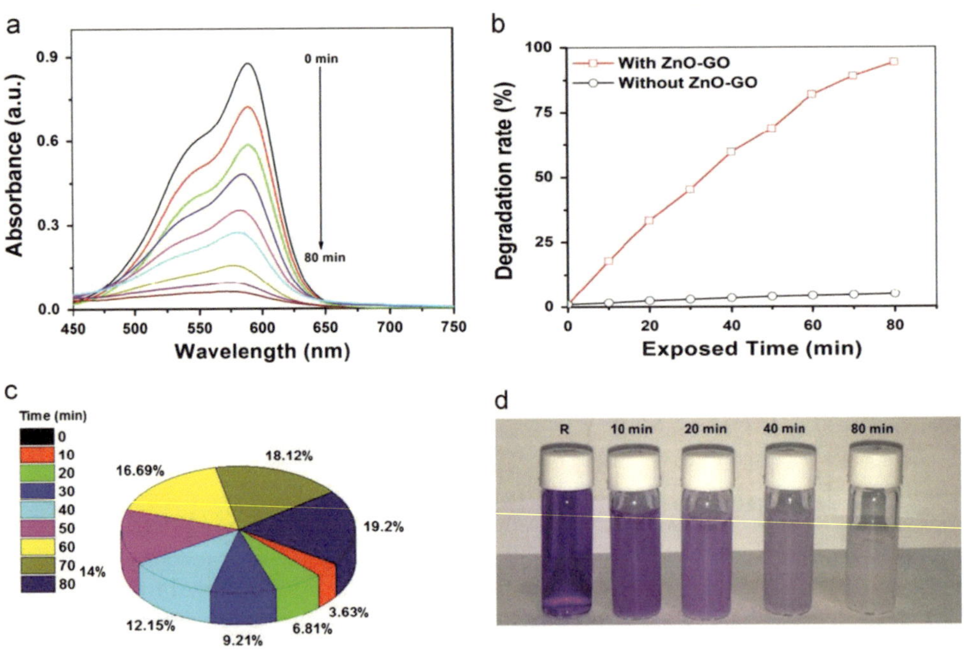

Figure 5.
(a) UV-vis absorbance spectra of decomposed Cv dye solution over ZnO-GO nanohybrid under light illumination, (b) the degradation rate (%) versus exposed time, (c) Cv dye degradation pie chart as a function of exposed time, and (d) the photographs of decomposed Cv dye solution before and after the photocatalytic reaction [17].

over ZnO-GO nanohybrid photocatalyst was attributed to GO as supporting material which enhanced the absorption capacity and suppressed the recombination of photo-generated charge carriers [17]. According to another study, GO was electrophoretically deposited on fluorinated tin oxide at 50 V in 30 s, and the GO deposited specimens were further engrafted by the chemically synthesized ZnO quantum dots using spin-coating technique. The electrode thus prepared, which successfully sensed ethyl acetate with substantially high sensitivity of ~16.035 mA mM^{-1} cm^{-2} [18]. Graphene was also be used in the modification of buffer layer to fabricate small-molecule organic solar cells. After the addition of graphene, the solar cells exhibited an improved power conversion efficiency of ~3.63%. This improvement could be ascribed to the reduction in the series resistance as well as better interfacial contacts between the layers brought about by graphene [19].

Graphene being a two-dimensional material also acts as a solid lubricant and offers the unique properties against friction and wear which are not observed in other materials. Graphene exhibits impressive tribological properties owing to its high chemical stability, mechanical durability and easy shear capability of its densely packed and atomically smooth surface. Since, it is ultrathin even with multilayers, it could be applied to nanoscale or microscale systems such as microelectromechanical systems (MEMS) and nanoelectromechnical systems (NEMS) with oscillating, rotating, and sliding contacts to reduce stiction, friction, and wear [20].

The studies illustrated above demonstrate the potential of graphene as an emerging material applicable in almost all the developing fields, and hence, it has attracted pivotal importance in the new age technology and innovation. Through this, this chapter exhibits the vast ambits of graphene with a hope to attract scientists and engineers to contribute into its challenging domains.

Author details

Sadia Ameen[1*], Rishikesh Godbole[2], M. Shaheer Akhtar[3] and Hyung-Shik Shin[2,4]

1 Advanced Materials and Devices Laboratory, Department of Bio-Convergence Science, Jeongeup Campus, Jeonbuk National University, Republic of Korea

2 Energy Materials and Surface Science Laboratory, Solar Energy Research Center, School of Chemical Engineering, Jeonbuk National University, Jeonju, Republic of Korea

3 New and Renewable Energy Material Development Center (NewREC), Jeonbuk National University, Jeonbuk, Republic of Korea

4 Korea Basic Science Institute (KBSI), Daejeon, Republic of Korea

*Address all correspondence to: sadiaameen@jbnu.ac.kr

References

[1] Du X, Skachko I, Barker A, Andrei EY. Approaching ballistic transport in suspended graphene. Nature Nanotechnology. 2008;**3**:491-495

[2] Park S, Ruoff RS. Chemical methods for the production of graphenes. Nature Nanotechnology. 2009;**4**:217-224

[3] Stoller MD, Park S, Zhu Y, An J, Ruoff RS. Graphene-based ultracapacitors. Nano Letters. 2008;**8**:3498-3502

[4] Zhang LL, Zhou R, Zhao XS. Graphene-based materials as supercapacitor electrodes. Journal of Materials Chemistry. 2010;**20**:5983-5992

[5] Berger C, Song Z, Li T, Li X, Ogbazghi AY, Feng R, et al. Ultrathin epitaxial graphite: 2D electron gas properties and a route toward graphene-based nanoelectronics. The Journal of Physical Chemistry. B. 2004;**108**:19912-19916

[6] Wang Z, Puls CP, Staley NE, Zhang Y, Todd A, Xu J, et al. Technology ready use of single layer graphene as a transparent electrode for hybrid photovoltaic devices. Physica E. 2011;**44**:521-524

[7] Wang X, Zhi L, Mullen K. Transparent, conductive graphene electrodes for dye-sensitized solar cells. Nano Letters. 2008;**8**:323-327

[8] Jeong JH, Jung DW, Kong BS, Shin CM, Oh ES. The effect of graphene nanosheets as an additive for anode materials in lithium ion batteries. Korean Journal of Chemical Engineering. 2011;**28**:2202-2205

[9] Fowler JD, Allen MJ, Tung VC, Yang Y, Kaner RB, Weiller BH. Practical chemical sensors from chemically derived graphene. ACS Nano. 2009;**3**:301-306

[10] Wang ZG, Chen YF, Li PJ, Hao X, Fu Y, Chen K, et al. Effects of methane flux on structural and transport properties of CVD-grown graphene films. Vacuum. 2012;**86**:895-898

[11] Mendoza F, Limbu TB, Weiner BR, Morell G. Hot Filament Chemical Vapor Deposition: Enabling the Scalable Synthesis of Bilayer Graphene and Other Carbon Materials. IntechOpen; 2016. DOI: 10.5772/63921

[12] Seo HK, Song M, Ameen S, Akhtar MS, Shin HS. New counter electrode of hot filament chemical vapor deposited graphene thin film for dye sensitized solar cell. Chemical Engineering Journal. 2013;**222**:464-471

[13] Song M, Seo HK, Ameen S, Akhtar MS, Shin HS. Low resistance transparent graphene-like carbon thin film substrates for high performance dye sensitized solar cells. Electrochimica Acta. 2014;**115**:559-565

[14] Song M, Ameen S, Akhtar MS, Seo HK, Shin HS. HFCVD grown graphene like carbon–nickel nanocomposite thin film as effective counter electrode for dye sensitized solar cells. Materials Research Bulletin. 2013;**48**:4538-4543

[15] Ameen S, Seo HK, Akhtar MS, Shin HS. Novel graphene/polyaniline nanocomposites and its photocatalytic activity toward the degradation of rose Bengal dye. Chemical Engineering Journal. 2012;**210**:220-228

[16] Ameen S, Akhtar MS, Shin HS. Hydrazine chemical sensing by modified electrode based on in situ electrochemically synthesized

polyaniline/graphene composite thin film. Sensors and Actuators B: Chemical. 2012;**173**:177-183

[17] Ameen S, Akhtar MS, Seo HK, Shin HS. Advanced ZnO–graphene oxide nanohybrid and its photocatalytic Applications. Materials Letters. 2013;**100**:261-265

[18] Ameen S, Akhtar MS, Seo HK, Shin HS. ZnO quantum dots engrafted graphene oxide thin film electrode for low level detection of ethyl acetate. Materials Letters. 2014;**136**:379-383

[19] Nazim M, Ameen S, Akhtar MS, Shin HK. D-π-A-π-D type thiazolo [5,4-d]thiazole-core organic chromophore and graphene modified PEDOT:PSS buffer layer for efficient bulk heterojunction organic solar cells. Solar Energy. 2018;**171**:366-373

[20] Berman D, Erdemir A, Sumant A. Graphene: A new emerging lubricant. Materials Today. 2014;**17**:31-42

Chapter 2

Graphene: Fabrication Methods, Properties, and Applications in Modern Industries

Ramesh Rudrapati

Abstract

Graphene research has fast-tracked exponentially since 2004 when graphene was isolated and characterized by Scotch Tape method by Geim and Novoselov and found unique electronic properties in it. Graphene is considered a promising material for industrial application based on the intensive laboratory-scale research in the fields of physics, chemistry, materials science, and engineering, over the last decade. The number of academic research publications related to various aspects of production, material properties, and applications of graphene has got increased substantially. With such a massive curiosity in graphene, it is imperious for both experts and the layman to keep up with both current graphene technology and the history of graphene technology. In the present study, focus has been given to addresses the disseminating graphene research with production, properties, and applicatory approach. The concluding remarks have been drawn from the present work.

Keywords: graphene, graphene production, graphene applications, graphene properties

1. Introduction

Materials research is engulfed globally to create products to solve real world applications. Major focus has given to make ultra-thin carbon films. Among the others, graphene is thinnest form of smart material. Graphene materials were discovered by Andre Geim and Konstantin Novoselov at the University of Manchester and their excellent contributions were recognized with Nobel prize in 2010. Since, graphene was discovered in 2004, it becomes most popular for human civilization' thus, it is one of the most wonderful achievements of science and technology [1]. Graphene is one of the high stable materials, because of very close or tight packing of atoms in the crystal lattice of graphene and its related materials [2, 3]. Due to enormous reputation of graphene, large research communities, had given significant attention to design and development of graphene materials for varies industrial applications like longer-lasting batteries, efficient solar cells, corrosion prevention, circuit boards, display panels, medical purposes, [4] etc. Basic structure of graphene is a two-dimensional single layer of sp^2 bonded carbon atoms systematically formed in a hexagonal lattice. Schematic structure of graphene is clearly shown in **Figure 1**. Graphene is a building block for all graphitic materials as it can be made into 0-dimensional fullerenes, 1-dimensional carbon nano tubes and 3-dimensional graphite structures. Graphene and its types are shown in **Figure 2**.

Graphene is considered as smart materials which possesses excellent mechanical, electronic, thermal, barrier, optical and chemical properties such as high surface area, superior thermal conductivity, high electron movement, high young's modulus, excellent high light transmittance, chemically stable, high level of transparent, [5–7] etc. Because of all stated properties, graphene has potential materials which suitable for variety of advanced/smart industrial applications like medical, paper, electronics and many others [8, 9]. Excellent properties of graphene those are well suited for interdisciplinary applications, giving strong

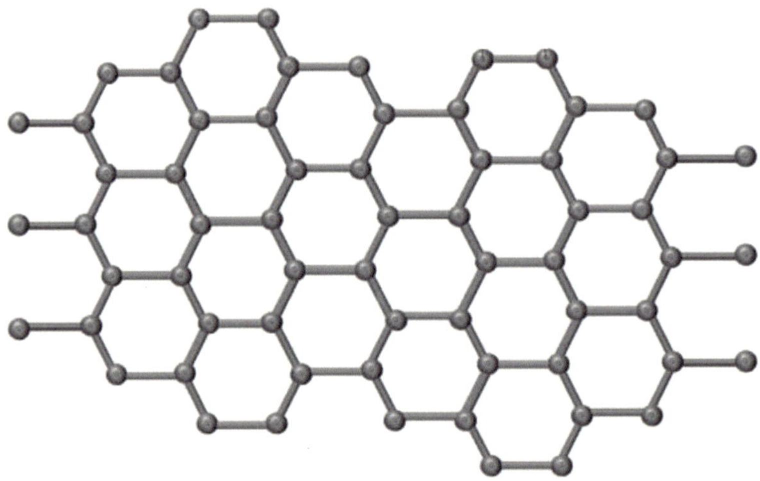

Figure 1.
Basic structure of graphene [6].

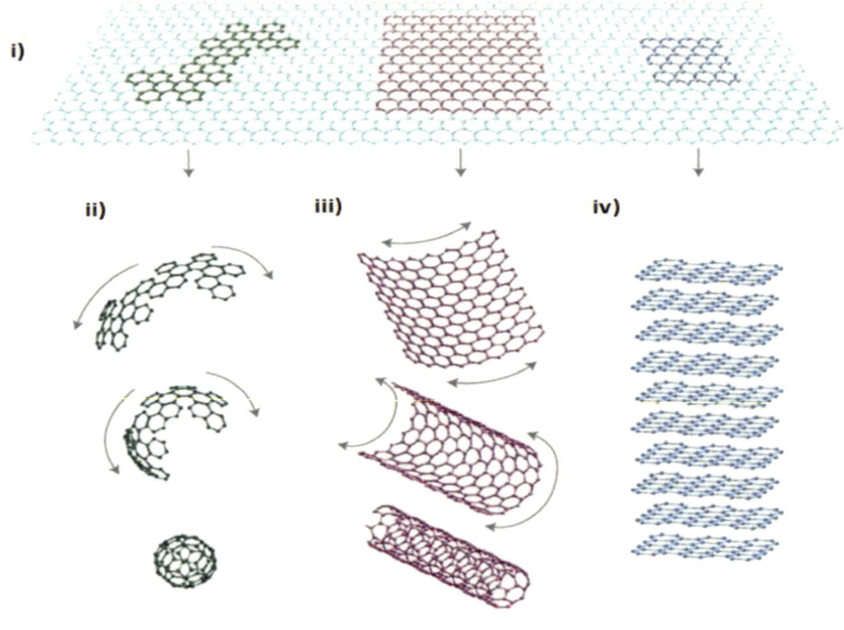

Figure 2.
(i) Graphene and its related structures: (ii) fullerene; (iii) carbon nanotubes; and (iv) graphite [15].

indication for scientists among the physics, chemistry, materials and metallurgy, biology, bio medical, electronics, energy sectors and many other allied fields for mass production of graphene [10–14]. From the buzz of many research communities around the globe, it is mentioned that start-up companies and existing industries have been initiated to use graphene for producing graphene related products.

Various fabrication methods for production of graphene and its related parts are developed but integrating graphene with other materials/products is challenging task. Great open for syntheses of graphene for producing high quality pats with advancements in existing methods. Usage of graphene in various industrial applications is in initial stages, it needs to commercialize rapidly. If commercial fabrication methods are developed, then price ranges of graphene-based products may go down. Graphene has entered into new era in the development field to use in different applications from medical to space to naval. The challenges are enchanting and can showcase scientifically better properties which well suits for various applications. Schematic diagram showing graphene production techniques, properties and application are given in **Figure 3**.

With increasing demands intelligent materials to meet needs of modern world. Graphene is one such material among the others. Several investigators have been given attention to various aspects related to design and development of graphene [17–19]. Number of research papers reported to literature per year is given in **Figure 4**. Now, recent advancements related to graphene research are needed for academicians, scientists, industrialists and investigators. By considering this aspect, present study is planned to discuss recent progress in material engineering related

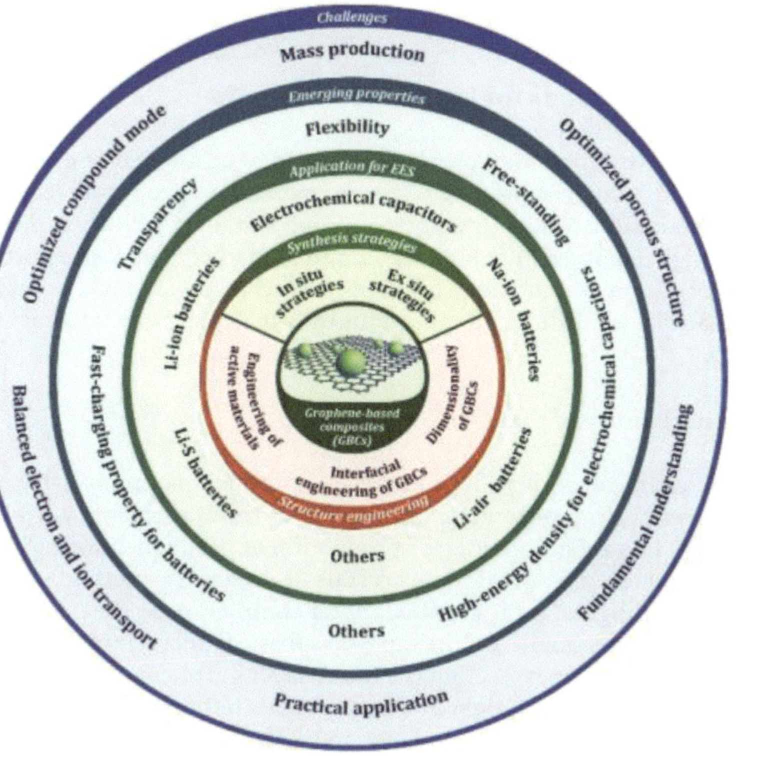

Figure 3.
Schematic illustration related to production, properties and applications of graphene-based composites (GBC) [16].

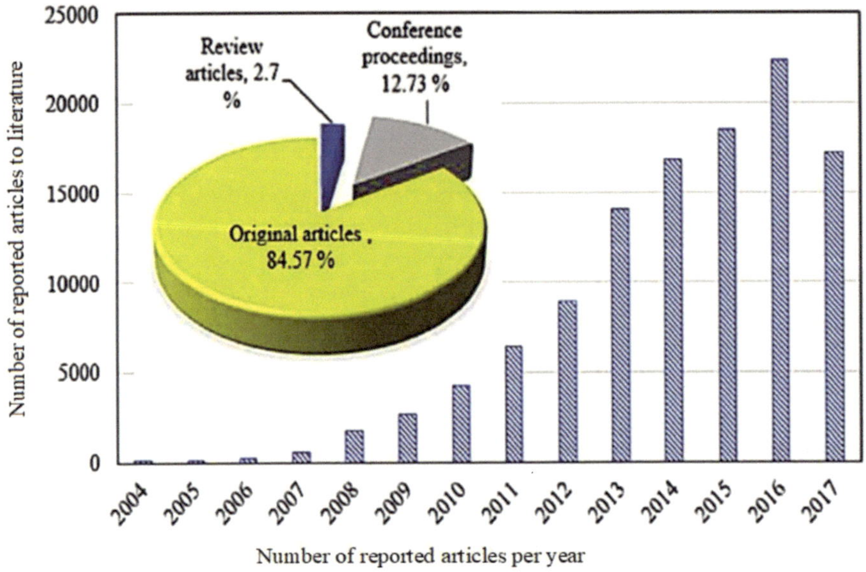

Figure 4.
Status of research publications reported corresponding year [20].

to graphene covering graphene fabrication methods, properties and applications. It is difficult to cover every application of graphene, but, some important aspects of graphene like basic introduction/definition, properties and applications of graphene have been presented.

2. Fabrication methods for graphene

As already mentioned, that graphene is found to be intelligent material for many advanced industrial purposes despite its excellent properties. Since graphene was discovered, industrialists have been finding suitable fabrication methods for producing high quality, defect free, stable and high yield and cost-effective methodologies. Fabrication methodology of graphene is challenging task, because utilization this material for different applications mostly depends on fabrication methods at large scale.

2.1 Micromechanical exfoliation

Micromechanical exfoliation is method of producing graphene-based materials which involves peeling systematically ordered pyrolytic graphite with the use of adhesive tape. It is a methodology of production of graphene, during this process graphene is separated from graphite crystals Peeling is the method used to produce graphene by peeling it off the graphite. After completion of the peeling, multilayer graphene's are remains on the tape. Graphene is sliced into various flakes of few layers by continuously peeling the multi-layer graphene. In this process layers of graphene are bonded strongly by van der Waals bonding. Schematic diagram of micromechanical exfoliation is shown in **Figure 5** [21]. It is simple easy manufacturing method for producing graphene materials but, it is not suitable for large scale growth of graphene materials. Information related to production procedure to make graphene can available in literature [22, 23].

2.2 Liquid-phase exfoliation (LPE)

Liquid phase exfoliation is method of production of graphene materials by using solvent like acetic acid, sulfuric acid and hydrogen peroxide, to exfoliate graphite through ultrasonication. Sonication methodology is used in LPE to exfoliate the graphene from graphite material, as graphite contains different layers of graphene which attached by Van der Waals forces. This method used to create graphene nanoribbons, but large-scale growth of graphene is difficult task in this method also. Schematic representation of LPE process is shown in **Figure 6**. Details of LPE to produce graphene can get from literature [24, 25].

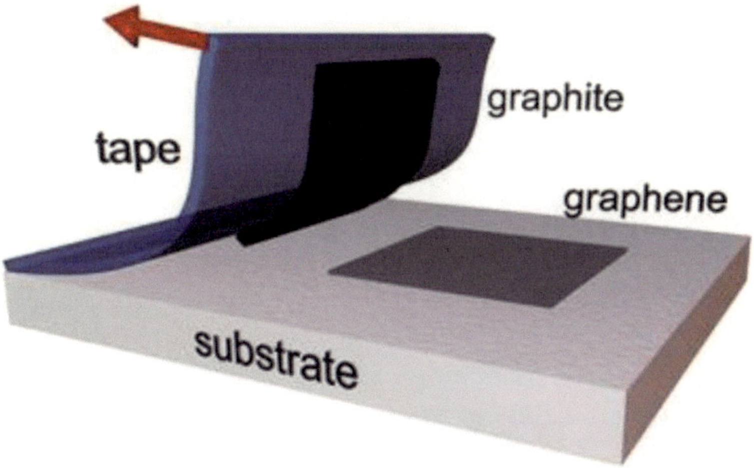

Figure 5.
Working procedure of micromechanical exfoliation process.

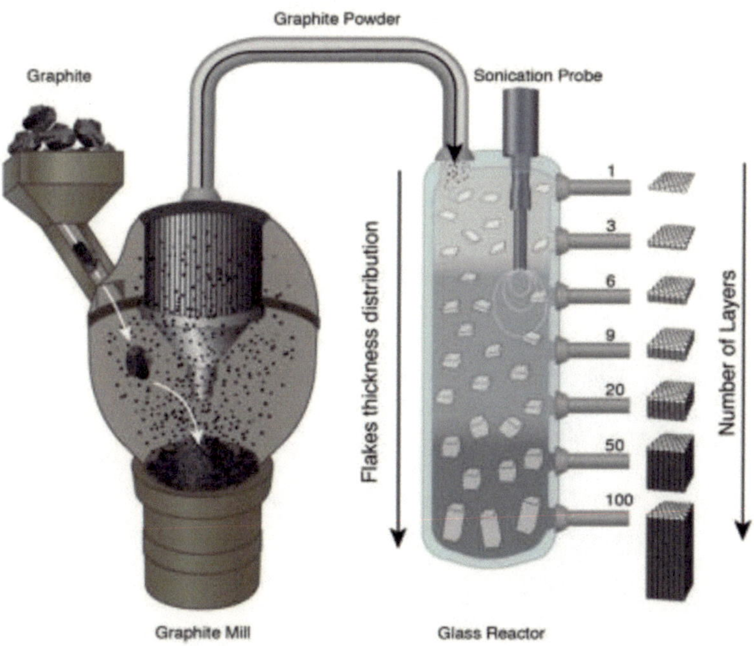

Figure 6.
Schematic diagram of production of graphene in LPE [26].

2.3 Chemical vapor deposition (CVD)

CVD is one of the important deposition methodologies used to transition metals. In CVD process, nickel and copper used for large scale production of graphene. During CVD process, film of metallic catalyst deposits on the substrate. Chemical etching is performed on the deposited material on the substrate. After chemical etching, a mixture which containing the carbon is passed into the reaction chamber. Experimental set up of CVD process is shown in **Figure 7**. The quality of the

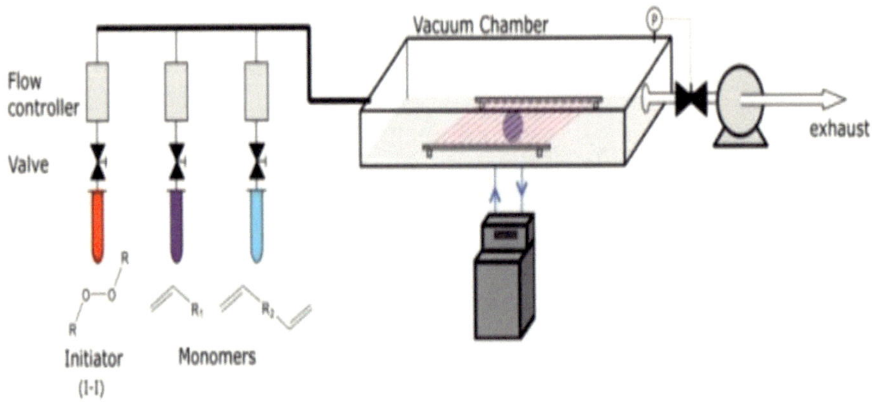

Figure 7.
Schematic diagram of experimental set up of CVD process [29].

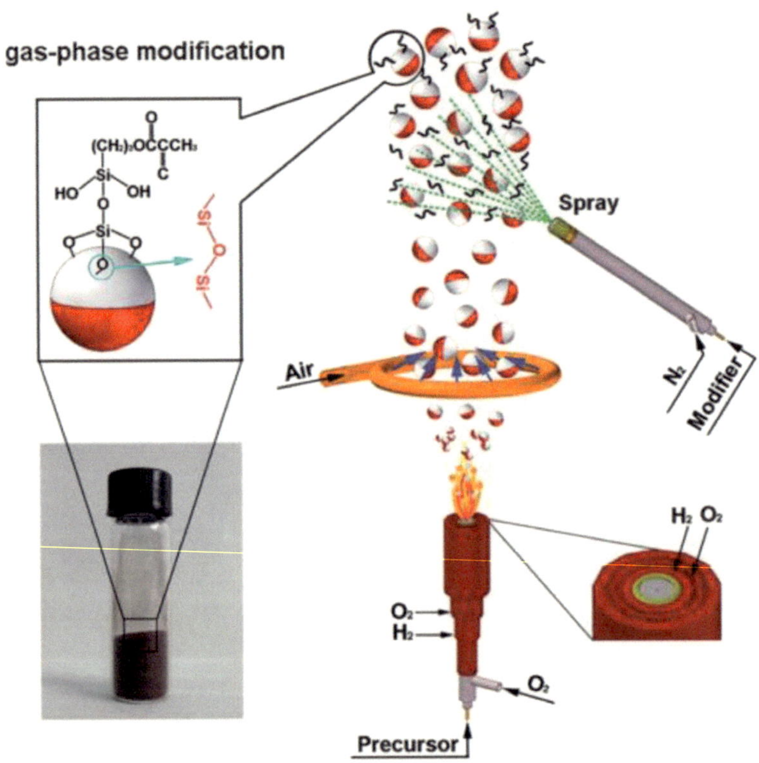

Figure 8.
Experimental set up of production of graphene in flame synthesis.

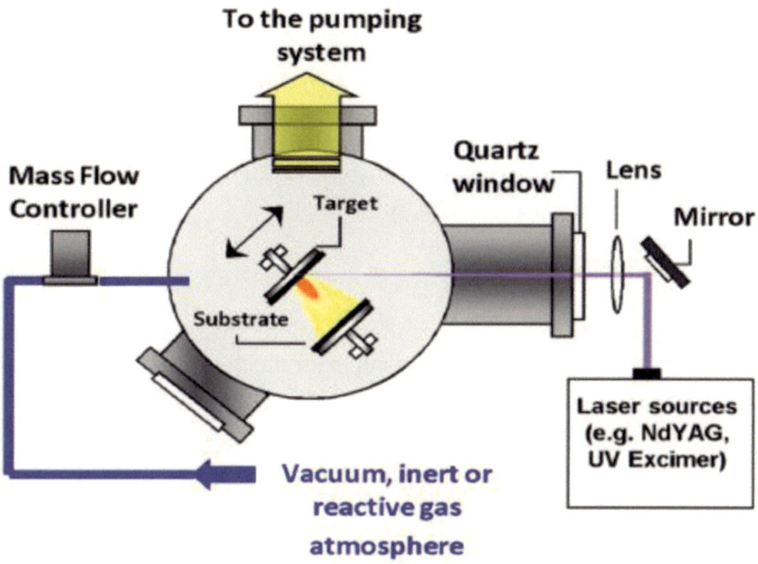

Figure 9.
Working of PLD of graphene [35].

graphene obtained from CVD process is high quality. More information regarding to CVD can be available in literature [27, 28].

2.4 Flame synthesis

Flame synthesis is widely used mass production method for making nano particles. This method is not well adopted for production of graphene as compared to chemical vapor deposition. Some researchers have focused to use of flame synthesis for making graphene materials due to its advantages like scalability and cost effectiveness. Experimental procedure of flame synthesis is given in **Figure 8** [30]. Some researchers suggested that flame synthesis has potential to produce graphene economically [31, 32].

2.5 Pulsed laser deposition (PLD)

PLD is a widely used method of growth approach for producing almost all types of materials. During PLD process, laser energy source is outside the chamber; and chamber is maintained ultrahigh vacuum. Schematic diagram of PLD is given in **Figure 9**. In this process, material is deposited by at an angle of 45° by stoichiometry transfer between ablated target and substrate material. During this process, substrates are added to its surfaces parallel to the target at distance of 2–10 mm. Main advantages of PLD process is low temperature growth rate achieved such that high-quality graphene made without defects. Reader can get more details related PLD in literature [33, 34].

3. Properties of graphene

The rapid interest in graphene has started due its unusual properties exhibits/ possesses by graphene materials. Excellent properties of graphene fascinate to be huge potential for various applications. Properties of graphene which reported to literature are mostly single layer defect free. Some of the details about the properties of graphene are discussed in following paragraph.

3.1 Electronic properties

The graphene revolution has started with invention of excellent electrical and electronics properties. The properties of graphene materials are highly depending on number of layers used to produce graphene sheets. It is zero overlap semimetal and it possesses higher electrical conductivity. Graphene is highly suitable for transistors applications due to electron-hole effect.

3.2 Mechanical properties

Graphene and its related materials exhibit excellent mechanical properties. Graphene is strongest material, because of superior mechanical properties of graphene. It is important note that mechanical properties are depends on purity of graphene sheets.

3.3 Optical properties

Optical properties of graphene are highly related to the electronics properties. As mentioned earlier that graphene exhibits excellent electronics properties and this phenomenon indicates graphene possesses better optical properties. Graphene is at can most full transparent materials as it can absorb 2.3% fraction of light [12]. Because of graphene materials exhibits behaviour which enable graphene to extraordinary optical properties. One can get more information related optical properties of graphene from literature [36].

3.4 Thermal properties

Thermal conductivity of graphene is depending on the diffusive and ballistic conditions at higher and lower temperature ranges respectively. Better thermal conductivity of graphene materials is highly depending on quality of graphene sheets. More details of thermal properties of graphene are available from literature [37].

3.5 Chemical properties

In chemical reaction point of view pure form of graphene is mostly not reactive. Chemical properties of graphene are critically influenced by its surface characteristics and thickness of graphene layers. Single layer graphene materials are highly chemically reactive then the multi-layer graphene materials. Reactiveness of graphene materials are controlled by nitrene chemistry methodology [38, 39].

4. Applications of graphene

Graphene is intelligent material which exhibits excellent properties used for various industrial applications. Some of the notable applications where graphene started using to create the parts are given as follows.

4.1 Graphene in high speed electronics

One can found the importance of electronics in most of the industrial applications from medical to mechanical to optical to energy. Conductivity property of electronics need to high for making electronics devices efficient and effective for usage in real world applications. Graphene is one of the advanced materials that exhibits high conductivity which considered as ideal for high speed electronics.

However, commercial applications of graphene have in initial stages only. Graphene is zero band gap material, more studies needed for of usage industrial applications. Research groups indicating that graphene field is advancing fast to create high speed graphene transistors for applying consumer electronic devices very soon.

4.2 Graphene in data storage

Data storage is one of the important areas of research. Investigators are developing the powerful small size hard drives to store higher capacity of data. Researchers suggested that replacing indium tin oxide electrodes with polymers and graphene oxide exhibit write-read-read-rewrite features. Graphene based storage devices are 10 times more powerful than the currently available storage drives. Making small size storage devices is not an issue but increasing capacity levels of storage devices is much importance task. With applications of graphene oxide devices will create big difference in modern industrial environments.

4.3 Graphene in LCD smart windows and OLED displays

Liquid crystal display (LCD) smart window is flexible device which consist of a layer of liquid crystals sandwiched between two flexible electrodes made of flexible polymer and graphene. Organic light emitting diode (OLED) windows are also utilized graphene-based OLED counter electrodes. Currently LCDs and OLED technologies utilize indium tin oxide counter electrodes. These materials are brittle in nature and limited availability in the world. Compared to indium tin oxide, graphene is flexible and availability is more/limitless. Usage of graphene in producing flexible smart devices like mobiles and tablet devices is an important research area due to its excellent properties.

4.4 Graphene in supercapacitors

Present days electronics are occupying almost every industrial application. Energy storage devices are highly required in every electronics to delivering high electric currents within short time. Supercapacitor is one of the important energy storage devices which utilizes high internal surface area to store charge to delivers higher currents compared to normal capacitors. Graphene can be highly suitable for making supercapacitors due to its higher internal area property. More research attempts are performing to create graphene-based supercapacitors for many advanced applications.

4.5 Graphene in solar cells/photovoltaic cell

Solar energy is one of the alternatives, and usage is increasing due to shortage of fusil fuels. Solar cell is important element in solar device which plays critical role to absorb energy from sun light. Presently, platinum-based electronics are using to produce solar cells or photovoltaic cells. Due to higher cost of platinum based solar cells limits its usage in industries. On the other hand, graphene is excellent conductor which is potential material for solar cells. Graphene based electrodes can be made low cost as well as weight while maintaining the efficiency [14].

4.6 Graphene in thermoelectric applications

Thermoelectric materials (TEM) are useful to convert thermal energy to electrical energy and vice versa. TEM are highly used in Peltier coolers and thermoelectric power generators. Graphene materials exhibits excellent thermoelectric properties which triggering industrialists to make attention to use graphene in thermoelectric applications.

4.7 Graphene in shape memory materials

Shape memory polymers (SMP) are smart materials which have wide range of industrial applications from biomedical to space applications. Graphene is better alternative as shape memory material due to its shape memory, thermal and mechanical properties as compared to existing SMP material namely polyurethane.

4.8 Graphene in self-healing materials

Material properties like long term stability and durability are much needed for structural and coating applications. Presently, polymeric composites are using said purpose and efficiency of these materials depends on many interacting parameters such as environmental condition, erosion, corrosion, etc. Self-healing of material indicates that material should heal basically mechanical properties when material gets damaged. Graphene is found to be potential material among the other materials like polymer composites, metals, ceramics and its related alloys, due to shape memory effect and self-healing ability.

4.9 Multifunctional graphene nanocomposite foams for space applications

Space and aerospace are highly advanced industries need intelligent materials which combine functionalities with low weight, minimized volume and cost effectiveness. Weight and volume of material used for space application are significantly influences cost of the satellite/space vehicles. Joule heating property of metallic parts requires additional cooling devices which adds weight and cost. Graphene is found to be better alternative material for space applications due to its superior electronic and thermal properties. With us of graphene as material for space applications, weight, volume and cost can be optimized and Joule heating can also be suppressing due ballistic electron transport property of graphene.

4.10 Graphene in electrorheology materials

Electrorheology (ER) materials are important smart material where rheological properties of material like viscosity, shear stress and dynamic modulus, can be reversibly transformed by the application of external electric field. ER materials M are widely using for producing damper systems, ER polishing, tactile displays, medical devices, robotic actuators, etc. Graphene is one material which is well suited additive for ER material due to its unique properties.

4.11 Other applications

As mentioned earlier that usage and growth of graphene and related materials are increasing enormously in advanced applications [40] like gene delivery and bio imaging [41], tissue engineering [42], graphene based metal air batteries [43], graphene LED bulbs [44], graphene antennas [45], graphene functional inks [46], graphene based fabric [47], etc.

5. Conclusions

Following summary points are drawn from the present study of status of graphene research related to fabrication methods, properties and applications:

i. Graphene and its related materials are futuristic one's which got enormous importance in modern world.

ii. Production/syntheses of graphene is challenging task.

iii. Some fabrication techniques of graphene are discussed which are using presently.

iv. Various properties of graphene are discussed.

v. Important industrial applications of graphene are addressed in the chapter.

vi. Challenging tasks of graphene production and applications are discussed.

vii. From the study, it is stated that graphene is smart material which is well suited for many advanced industrial applications; various aspects related to graphene production, properties and real-world applications need to explore further to make graphene more intelligent material.

Acknowledgements

The author acknowledges to President, Bule Hora University, Dean, College of Engineering & Technology, Bule Hora University, for providing necessary support for this work.

Conflict of interest

The author declaring no 'conflict of interest.

Author details

Ramesh Rudrapati
Industrial Engineering Department, Bule Hora University, Bule Hora, Ethiopia

*Address all correspondence to: rameshrudrapati@gmail.com

References

[1] Tiwari SK, Kumar V, Huczko A, Oraon R, Adhikari AD, Nayak G. Magical allotropes of carbon: Prospects and applications. Critical Reviews in Solid State and Materials Sciences. 2016;**41**(4):257-317

[2] Tiwari SK, Mishra RK, Ha SK, Huczko A. Evolution of graphene oxide and graphene: From imagination to industrialization. ChemNanoMat. 2018;**4**(7):598-620

[3] Santosh KT, Sumanta S, Nannan W, Andrzej H. Graphene research and their outputs: Status and prospect. Journal of Science: Advanced Materials and Devices. 2020;**5**(1):10-29

[4] Edward PR, Dale ACB, Craig EB. A decade of graphene research: Production, applications and outlook. Materials Today. 2014;**17**:426-432

[5] Leonel IS, Daniel AM, Tomba JP, Carmen CR. Optimizing graphene production in ultrasonic devices. Ultrasonics. 2020;**100**:105989

[6] Josphat P, Patrick G, Thad CM. General overview of graphene: Production, properties and application in polymer composites. Materials Science and Engineering B. 2017;**215**:9-28

[7] Wei J, Atif R, Vo T, Inam F. Graphene nanoplatelets in epoxy system: Dispersion, reaggregation, and mechanical properties of nanocomposites. Journal of Nanomaterials. 2015;**12**:1-12

[8] Stankovich S, Dikin DA, Dommett GHB, Kohlhaas KM, Zimney EJ, Stach EA, et al. Graphene-based composite materials. Nature. 2006;**442**:282-286

[9] Dreyer DR, Ruoff RS, Bielawski CW. From conception to realization: An historical account of graphene and some perspectives for its future. Angewandte Chemie International Edition. 2010;**49**:9336-9344

[10] Zhang LL, Zhou R, Zhao XS. Graphene-based materials as supercapacitor electrodes. Journal of Materials Chemistry. 2010;**20**:5983-5992

[11] Schwierz F. Industry-compatible graphene transistors. Nature. 2011;**472**:41-42

[12] Wei D, Kivioja J. Graphene for energy solutions and its industrialization. Nanoscale. 2013;**5**(21):10108-10126

[13] Neuberger N, Adidharma H, Fan M. Graphene: A review of applications in the petroleum industry. Engineering. 2018;**167**:152-159

[14] Mahmoudi T, Wang Y, Hahn YB. Graphene and its derivatives for solar cells application. Nano Energy. 2018;**47**:51-65

[15] Geim AK, Novoselov KS. The rise of graphene. Nature Materials. 2007;**6**:183-191

[16] Bo W, Tingting R, Yong C, Fan J, Li P, Yu Z, et al. Graphene-based composites for electrochemical energy storage. Energy Storage Materials. 2020;**24**:22-51

[17] Mueller T, Xia F, Avouris P. Graphene photodetectors for high-speed optical communications. Nature Photonics. 2010;**4**:297-301

[18] Koratkar N. Graphene in Composite Materials: Synthesis, Characterization and Applications. USA: DEStech Publications Inc; 2013

[19] Ghany NAA, Elsherif SA, Handal HT. Revolution of graphene for

different applications: State-of-the-art. Surfaces and Interfaces. 2017;**9**:93-106

[20] Meysam TK, Ajay K, Emad O, Chngsoo K, Pradeep R. Synthesis, characterization, and properties of graphene reinforced metal-matrix nanocomposites. Composites Part B Engineering. 2020;**183**:107664

[21] Diana CCF. Low dimensional materials: Synthesis, characterization and applications [PhD thesis]. Universita Dellacalabria, Italy; 2017

[22] Robert CS, James LS, Peter VC. Micromechanical exfoliation of graphene on the atomistic scale. Physical Chemistry Chemical Physics. 2019;**21**:5716

[23] Alexandre A, Samuel MV, Miguel CL, Eugenio C. Graphene related magnetic materials: Micromechanical exfoliation of 2D layered magnets based on bimetallic anilate complexes with inserted $[Fe^{III}(acac_2\text{-trien})]^+$ and $[Fe^{III}(sal_2\text{-trien})]^+$ molecules. Chemical Science. 2015;**6**:4665

[24] Xu C, Chenzhen Z, Rui H, Yanglong H. Liquid-phase exfoliation, functionalization and applications of graphene. Nanoscale. 2011;**3**:2118-2126

[25] Majid M. Liquid-phase exfoliation (LPE) of graphite towards graphene: An ab initio study. Journal of Molecular Liquids. 2017;**230**:461-472

[26] Alan PK, Andressa TS, Diego PP, Roshini CP, Ricardo B, Ricardo VBO, et al. The worldwide graphene flake production. Advanced Materials. 2018;**30**:1-6.1803784

[27] Mattevi C, Kim H, Chhowalla M. A review of chemical vapour deposition of graphene on copper. Journal of Materials Chemistry. 2011;**21**:3324-3334

[28] Piran RK, Caterina D, Bruno D, Damian G, Robert SW, Marie-Blandine M, et al. The parameter space of graphene chemical vapor deposition on polycrystalline Cu. Journal of Physical Chemistry C. 2012;**42**:22492-22501

[29] Available from: http://news.mit.edu/2015/explained-chemical-vapor-deposition-0619

[30] Yunfeng L, Yanjie H, Hao J, Chunzhong L. Double-faced g-Fe_2O_3||SiO_2 nanohybrids: Flame synthesis, in situ selective modification and highly interfacial activity. Nanoscale. 2013;**5**:5360-5367

[31] Kammler HK, Madler L, Pratsinis SE. Flame synthesis of nanoparticles. Chemical Engineering & Technology. 2001;**24**:583-596

[32] Nasir KM, Stephen DT, Jafar FAS, Hisato Y, Alem-Mar BG, Bernard HK, et al. Flame synthesis of graphene films in open environments. Carbon. 2011;**49**:5064-5070

[33] Tite T, Donnet C, Loir AS, Reynaud S, Michalon JY, Vocanson F, et al. Graphene-based textured surface by pulsed laser deposition as a robust platform for surface enhanced Raman scattering applications. Applied Physics Letters. 2014;**104**:041912

[34] Margus K, Artjom B, Tauno K, Tea A, Aarne K, Ahti N, et al. Highly sensitive NO_2 sensors by pulsed laser deposition on graphene. Applied Physics Letters. 2016;**109**:113108

[35] Brahim A, Nasir KM, Adnan A, Marwan KK. Recent progress in the growth and applications of graphene as a smart material: A review. Frontiers in Materials. 2015;**2**:1-19

[36] Bonaccorso F, Sun Z, Hasan T, Ferrari AC. Graphene photonics and optoelectronics. Nature Photonics. 2010;**4**:611-622

[37] Tian X, Itkis ME, Bekyarova EB, Haddon RC. Anisotropic thermal and electrical properties of thin thermal interface layers of graphite nano platelet-based composites. Scientific Reports. 2013;**3**:1-6

[38] He H, Gao C. General approach to individually dispersed, highly soluble, and conductive graphene nanosheets functionalized by nitrene chemistry. Chemistry of Materials. 2010;**22**:5054-5064

[39] Chen D, Tang L, Li J. Graphene-based materials in electrochemistry. Chemical Society Reviews. 2010;**39**:3157-3180

[40] Dhinakaran V, Lavanya M, Vigneswari K, Ravichandran M, Vijayakumar MD. Review on exploration of graphene in diverse applications and its future horizon. Materials Today: Proceedings. 2020. DOI: 10.1016/j.matpr.2019.12.369

[41] Sumit G, Vinayak S, Shilpa S. Graphene-based nanomaterials for drug delivery and tissue engineering. Journal of Controlled Release. 2014;**173**:75-88

[42] Shin SR, Yi CL, Hae LJ, Parastoo K, Mohsen A, Yu AN, et al. Graphene-based materials for tissue engineering. Advanced Drug Delivery Reviews. 2016;**105**:255-274

[43] Eunjoo Y, Haoshen Z. Li–air rechargeable battery based on metal-free graphene nanosheet catalysts. ACS Nano. 2011;**5**(4):3020-3026

[44] Chung PL. U.S. Patent 9,933,121; 2018

[45] Zheyu F, Zheng L, Yumin W, Pulickel MA, Peter N, Naomi JH. Graphene–antenna sandwich photodeterctor. Nano Letters. 2012;**12**(7):3808-3813

[46] Kun F, Yibo W, Chaoyi Y, Yonggang Y, Yanan C, Jiaqi D, et al. Graphene oxide-based electrode inks for 3D-printed lithium-ion batteries. Advanced Materials. 2016;**28**(13):2587-2594

[47] Jiesheng R, Chaoxia W, Xuan Z, Tian C, Kunlin C, Yunjie Y, et al. Environmentally-friendly conductive cotton fabric as flexible strain sensor based on hot press reduced graphene oxide. Carbon. 2017;**111**:622-630

Chapter 3

Graphite Oxide: A Simple and Reproducible Synthesis Route

Ernesto Hernández-Hernández, Perla J. Hernández-Belmares, Mónica A. Ceniceros-Reyes, Oliverio S. Rodríguez-Fernández and Pablo González-Morones

Abstract

The synthesis of graphite oxide (GrO) by oxidation of graphite has been carried out by different procedures. In this chapter, we describe a simple synthesis route based on Hummers' method without the usage of $NaNO_3$ achieving nearly the same outcomes, and this methodology is directed toward high-quality scale production of GrO with similar properties compared with GrO obtained with traditional and improved Hummers' methods. The GrO was obtained in a series of batch reactions and characterized by different techniques, and the results showed identical interlayer *d-space*, type and content of oxygen functionalities, and I_D/I_G ratio. The high reproducibility of this methodology offers an efficient alternative for the large-scale production of graphene oxide.

Keywords: graphite oxide, graphene oxide, modified Hummers' method, reproducibility, oxidation process

1. Introduction

Graphenic materials have been one of the most studied materials in the history of humanity due to their outstanding properties such high thermal, electrical, mechanical, and permeability properties, among others [1]. For this reason, many potential applications have been proposed and demonstrated in scientific reports and patents. It has even been estimated that the global graphene market size will increase up to 38% from the years 2017 to 2025 [2], taking into an account its potential use in applications as automotive lightweight materials, aeronautics and energy, Li batteries, paints, functional coatings, solar cells, biosensors, membranes, and electronics, just to mention some of them [3–7]. One of the main technological challenges that engineers and scientific community face is the lack of new methods of large-scale production of graphene and its derivative. The graphite is inexpensive and available in large quantity and unfortunately does not readily exfoliate to yield individual graphene sheets. Graphite oxide (GrO) is a layered material produced by the oxidation of graphite. In contrast to pristine graphite, the GrO sheets, known as graphene oxide, are heavily oxygenated, bearing hydroxyl and epoxide functional groups on their basal planes, in addition to carbonyl and carboxyl groups presumable located at the sheet edges, nevertheless, there are certain features that still remain unknown among which stands out the chemical structure [8]. Particularly,

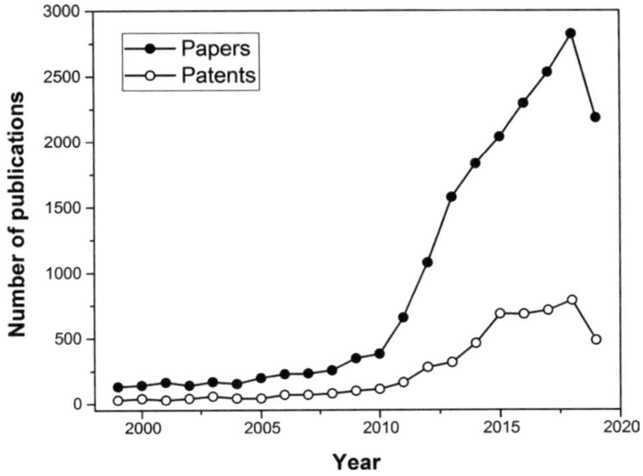

Figure 1.
Number of scientific articles and patents published related to the synthesis of graphite oxide. Source: Data obtained by the analysis using SciFinder.

graphene oxide (GO) has gained interest since it can be used for a wide scale of chemical transformations that include the reduction of graphene-like materials and its functionalization with other functional groups [9].

The number of publications in patents and research manuscripts related to the synthesis and production of graphite oxide and graphene oxide is shown in **Figure 1**. An abrupt increase in publications was observed after 2010. In 2018, approximately 2800 papers and 800 patents were published, and there is a tendency to increase the publications of both documents in the next years. This trend discloses the interest of these materials, which are expected to impact in the applications mentioned above.

The GrO can be prepared through several approaches, and each of them has their own advantages and flaws [10]. The main goal is to produce GrO at large scale with the best characteristics and high reproducibility; thus, the methodology here described consists in a variation of Hummers' method with important improvements that allow a successful synthesis of GrO.

2. Synthetic approaches

Graphene oxide can be synthesized via chemical oxidation of graphite, predominantly. Nevertheless, there are a few reports with an alternative electrochemical oxidation [11, 12]. Brodie's method, reported in 1859, was the first one in utilizing potassium chlorate to the mixture of graphite and nitric acid as the oxidant and intercalating agents, respectively. However, this technique has important flaws, such as the reaction time is about 4 days, low yield of the GO, the evolution of toxic acid vapors and NO_2/N_2O_4 gases, and the generation of highly explosive ClO_2 when chlorate mixed with strong acids [13]. Nearly 40 years later, Staudenmaier proposed the use of H_2SO_4 with HNO_3, but the explosive ClO_2 gas still remained as long as the prolonged reaction time. Based on Staudenmaier's work, Hummers and Offeman developed an alternative method that has been widely used for the synthesis of graphite oxide [14]. The chemicals used in this case were H_2SO_4 to intercalate graphite with the assistance of $NaNO_3$ and $KMnO_4$ as oxidant agents. The main reasons that this procedure is a reference in this matter are the use of $KMnO_4$ (strong oxidant) guarantees the completion of reaction within several hours, and the safety issue, in which there is no production of explosive ClO_2 due to the absence of $KClO_3$, and there is no generation of acid fog due to the replacement of HNO_3 with $NaNO_3$. Despite of its high efficiency and the safety matter, it still has some drawbacks: (1) the toxic gas generation (NO_2/N_2O_4), (2) residual Na^+ and NO_3^- ions are difficult to be removed after GrO synthesis and purification, and (3) incomplete oxidation resulting in the formation of graphite/GrO mixture [15, 16]. These problems have led to made several modifications to Hummers' method, and the main strategies includes the removal of $NaNO_3$. One of them is reported by Tour et al. [15] by increasing the amount of $KMnO_4$ and a 9:1 mixture of concentrated H_2SO_4/H_3PO_4 with a reaction time higher than 12 h. The GrO obtained by this methodology was highly oxidized with fewer defects in the basal plane and higher yield (77%), compared to GrO prepared by Hummers' method (40%). Shi et al. [16] removed the $NaNO_3$ from traditional Hummers method; with this simple modification, it was possible to produce GrO without affecting the yield and still had a high C/O ratio (2.36). Yu et al. [10] also reported a further improvement for $NaNO_3$-free Hummers' method by partly replacing $KMnO_4$ with K_2FeO_4; in addition, the amount of sulfuric acid was considerably reduced. This procedure resulted in a high yield (84%) compared to the Hummers traditional method.

The synthesis yield is normally estimated considering the mass ratio of graphite and graphite oxide. Methods aforementioned reported high yield by increasing the amount of oxidant agent and/or reaction time or by adding another reactant (acid or intercalating agent). These modifications may imply important disadvantages such as high cost, poor scalability, and practical applications. Based on $NaNO_3$-free Hummers' method, authors of the present chapter propose some changes in order to obtain oxidized sheets but keeping the graphenic properties and also to get these two themes in a scalable way. **Figure 2** illustrates the methodology in which the mixture of graphite-H_2SO_4 was previously sonicated to improve the intercalation of

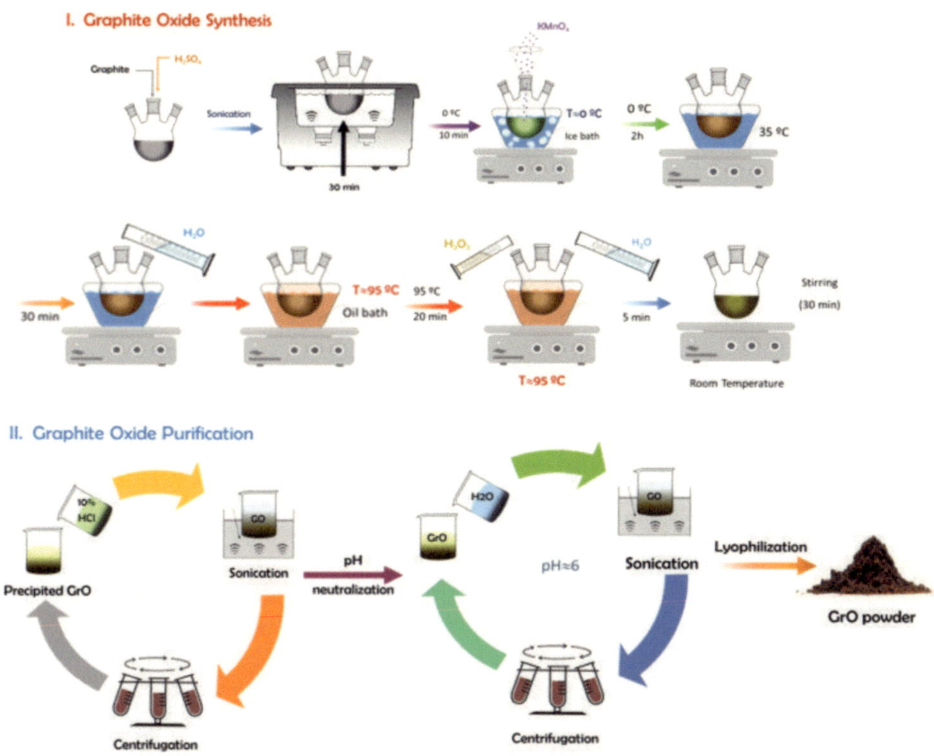

Figure 2.
Scheme of GrO synthesis proposed by the authors based on $NaNO_3$-free Hummers' method.

Method	Graphite (g)	Oxidant	Graphite/ oxidant ratio	Reaction time (h)	C/O atomic ratio	Yield of GrO (%)	Ref
Hummers	100	300 g $KMnO_4$ + 50 g $NaNO_3$	1:3	~2	2.1–2.9	40	[14]
Tour	3	18 g $KMnO_4$	1:6	>12	—	77	[15]
Shi	3	9 g $KMnO_4$	1:3	~2	2.36	—	[16]
Zhang	10	11 g $KMnO_4$ + 4 g $KFeO_4$	1:1.5	5	2.12	84	[10]
Author's work	2	6 g $KMnO_4$	1:3	~4	1.98–2.1	55	[17]

Table 1.
Comparison among different methods to synthesize GrO.

the acid between graphite galleries. Also, controlling the addition time of $KMnO_4$ ($t \leq 30$ min) and increasing the stirring time would enhance the diffusion of the $KMnO_4$ in the interlayer space. The obtained GrO in this procedure with the reaction time of 4 h has similar properties with those reported in the literature [17].

Table 1 shows noteworthy aspects of different methods to produce GrO, with respect to the chemicals involved, reaction time, C/O atomic ratio, and yield. The most important thing is the fact that has been proven to be a promising scalable method for obtaining graphite oxide, which was possible to demonstrate with the final features studied from several GrO samples synthesized.

3. GrO properties

In order to evaluate the reproducibility of the method proposed by the authors of the present chapter, GrO was synthesized in a total of 10 batch reactions, and the graphite oxide obtained was labeled GrO 1, GrO 2, ..., GrO n, where n corresponds to the reaction number. All GrO n samples were characterized by different analytical techniques and were compared to each other, including the precursor graphite, labeled as GT. The structural, chemical, thermal, and morphological properties are presented below.

3.1 Structural properties

The samples were analyzed by XRD to evaluate the crystalline structure of GT and different synthesized GrO. **Figure 3** shows the comparison among GT and three of the GrO samples (GrO 3, GrO 4, and GrO 7). XRD pattern of GT shows a characteristic diffraction peak (d_{002}) at 26.5° that corresponds to a d-space of 0.33 nm. After oxidation, this peak is no longer observable, instead a broad peak at a range of 11.1–11.6° can be assigned to d_{001}, which oscillates from 0.76 to 0.79 nm, and this increase in d spacing is attributed to the intercalation of water molecules and to the presence of the functional groups at the basal plane [8, 18, 19]. The XRD experimental data of each sample are presented in **Table 2**, and it is observed that the average position peak of all GrOs at $2\theta = 11.37 \pm 0.18°$ has an average interlayer space $d_{001} = 0.78 \pm 0.1$ nm. The Full Width at Half Maximum (FWHM) was used to estimate the thickness, L_c, by the Scherrer equation [20], whose results vary between 9.97 and 15 nm. The thickness, L_c, was used to calculate the average number of

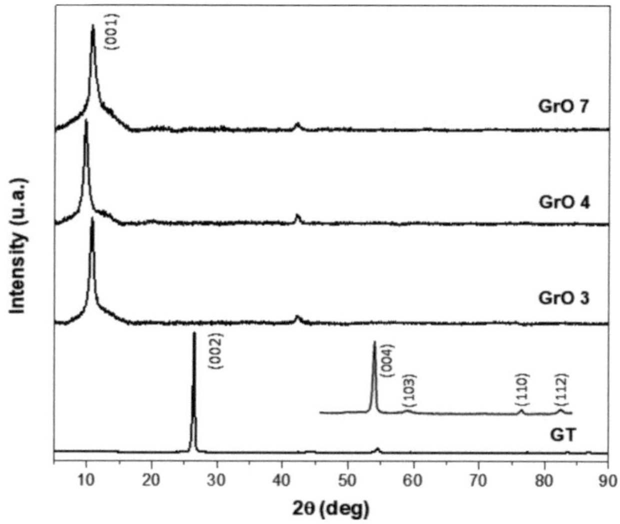

Figure 3.
X-ray diffraction patterns of GT, GrO 3, GrO 4, and GrO 7 [17].

Sample	2θ (°)	d_{001} (nm)	FWHM (°)	Thickness (nm)	Number of layers (thickness/d)
GT	26.50	0.34	0.32	25.17	74.90
GrO 1	11.37	0.78	0.70	11.46	14.75
GrO 2	11.43	0.77	0.55	14.54	18.81
GrO 3	11.15	0.79	0.61	12.99	16.38
GrO 4	11.27	0.78	0.53	15.00	19.12
GrO 5	11.19	0.79	0.70	11.47	14.52
GrO 6	11.15	0.79	0.61	13.09	16.50
GrO 7	11.64	0.76	0.80	9.97	13.13
GrO 8	11.61	0.76	0.59	13.44	17.66
GrO 9	11.37	0.78	0.56	14.37	18.48
GrO 10	11.49	0.77	0.61	13.19	17.15
Average	11.37	0.78	0.63	12.95	16.65
Std. dev.	0.18	0.01	0.08	1.58	2.00

Table 2.
X-ray data comparison among different GrO samples.

layers, being of 16.65 ± 2 for GrO, which is significantly low compared with the ~75 layers estimated for GT, and this indicates that periodic structure of graphite has been disrupt, and it has partially exfoliated forming small stacks of few layers.

The structural changes caused by oxidation process were also monitored by Raman spectroscopy. The Raman spectra of GT and GrO 2, GrO 4, and GrO 6 are compared in **Figure 4**. The GT exhibits a sharp and strong G band at around 1580 cm^{-1}, associated with bond stretching of the sp^2 carbon pairs in both rings and chains, and a weak and broad D band at 1350 cm^{-1}, associated with the presence of defects in graphite materials such as bond-angle disorder, bond-length disorder, vacancies, and etch defects [21]. The blue shift of the G band and the significantly

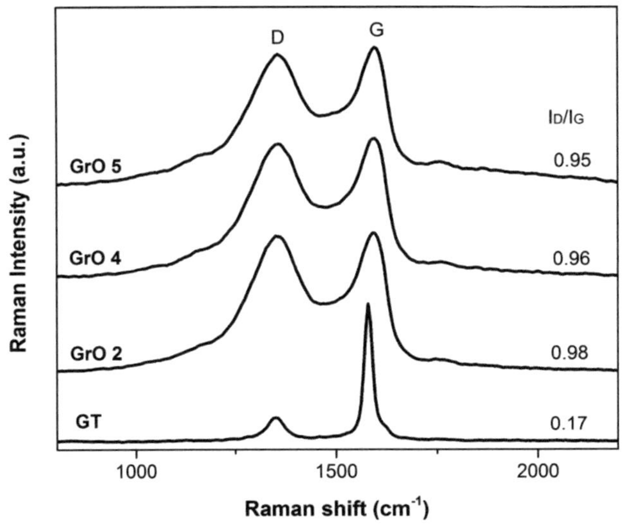

Figure 4.
Raman spectra of GT, GrO 2, GrO 4, and GrO 5 [17].

Sample	D band			G band			I_D/I_G	La (nm)
	FWHM	Position (cm^{-1})	Raman intensity	FWHM	Position (cm^{-1})	Raman intensity		
GT	57.75	1350.63	0.17	24.39	1579.25	1.00	0.17	110.14
GrO 1	199.13	1362.50	0.98	174.08	1596.59	1.00	0.98	19.67
GrO 2	188.79	1350.63	0.98	135.81	1590.81	1.00	0.98	19.65
GrO 3	195.98	1356.57	0.95	153.29	1596.59	1.00	0.95	20.27
GrO 4	206.51	1356.57	0.96	153.24	1596.59	1.00	0.96	20.04
GrO 5	191.20	1356.57	0.95	172.73	1596.59	1.00	0.95	20.25
GrO 6	190.69	1356.57	0.99	178.67	1596.59	1.00	0.99	19.41
GrO 7	144.54	1348.20	0.80	93.17	1588.67	0.86	0.94	20.53
GrO 8	163.68	1348.20	0.80	101.67	1588.67	0.86	0.93	20.67
GrO 9	173.00	1360.09	0.90	113.99	1594.45	0.93	0.96	19.98
GrO 10	169.57	1360.09	0.896	126.74	1600.23	0.946	0.95	20.29
						Average	0.96	20.08
						Std. dev.	0.02	0.41

Table 3.
Analysis of results obtained by Raman spectroscopy for GT and all GrO samples.

increase in the width and intensity of D band for all GrO samples with respect to GT can be associated with the defects induced in the hexagonal carbon network by the formation of oxygen functionalities and the parallel incorporation of sp^3 bonds during the oxidation of graphite (see **Table 3**). In addition, the notable increase in the intensity ratio I_D/I_G from 0.17 for GT to 0.96 ± 0.02 for GrO reveals a drastic decrease in the size of carbon sp^2 domains [22, 23] and can be corroborated by calculating the crystallite size, La, which is considerably less for GrO (~20 nm) than GT (110 nm) [24].

3.2 Thermal properties

Thermal stability of all GrO samples was evaluated by TGA, and some of them are presented in **Figure 5a** and were compared with GT that remains thermally stable to a temperature of above 700°C, whereas the thermal degradation of GrO presents several weight losses, the first at around 10% below 100°C is associated with the vaporization of adsorbed water molecules onto GO sheets, and the second weight loss of 30% is observed from 150 to 280°C, which is attributed to the thermal decomposition of labile oxygen functionalities, and it is also observed a small weight loss (~10%) from 270°C to 600°C, which is attributed to the removal of more thermally stable oxygen functional groups such as carbonyl groups [21, 25, 26]. The derivative weight loss curve of GrO presented in **Figure 5b** displays a maximum at 217–220°C related to the degradation of functional groups, and **Table 4** presents the weight loss at this temperature for all samples, whose average value is 30% with a degradation temperature of 218°C.

3.3 Chemical characterization

The FTIR and XPS analysis reveal significant chemical changes of GrO samples owing to oxidation process. FT-IR spectra in **Figure 6** compare results from GrO 6, GrO 7, and GrO 8 with GT. In all cases, GrO exhibited a broad peak at 3000–3700 cm^{-1}, which is attributed to O—H stretch vibration of hydroxyl, carboxyl, and intercalated free water. The vibrational peak at 1725 cm^{-1} is associated with the C=O stretch of both carboxyl and carbonyl groups, and the vibrational peak at 1623 cm^{-1} is assigned to the overlapped frequencies of bending modes of trapped water molecules and C=C stretch of unoxidized sp^2 carbon domains [27]. The O—H deformations of the C—OH groups appear at 1400 cm^{-1}. The peaks at 1220 cm^{-1} and 1057 cm^{-1} are associated with C—O stretching of epoxy and alkoxy groups, respectively [28].

On the other hand, the elemental chemical information of GrO 2 and GrO 7 samples was obtained by XPS analysis. The XPS survey spectra presented in **Figure 7** show the C 1s peak at 284.4 eV and O 1s peak at 533.5 eV, with an atomic content of oxygen of 34.6% and 32.1% for GrO 2 and GrO 7, respectively, being similar to that calculated by TGA. The atomic ratio C/O was 1.89 for GrO 2 and 2.08 for GrO 7, and these values coincide with those obtained for different GrOs synthesized by other techniques, as described in **Table 1**.

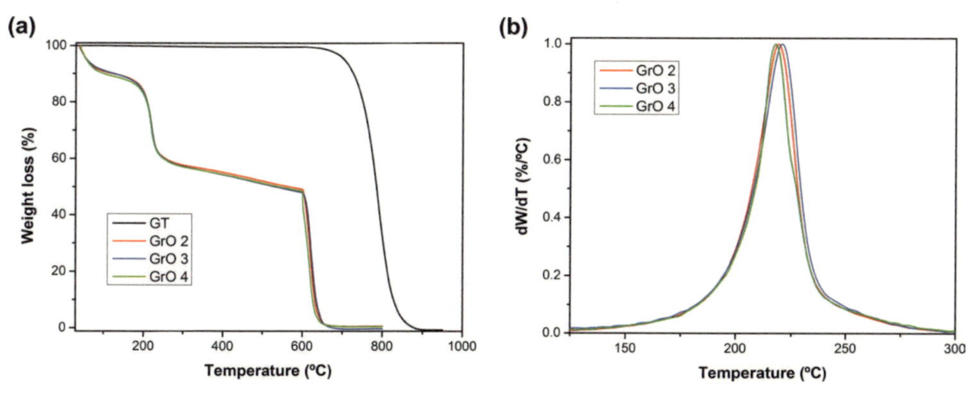

Figure 5.
Thermal behavior of GT and GrO: (a) TGA curves and (b) dW/dT curves of GT, GrO 2, GrO 3, and GrO 4 [17].

Sample	% Weight loss at 150–280°C	T_{max} (°C)
GrO 1	28.77	219.86
GrO 2	30.06	218.94
GrO 3	30.91	220.63
GrO 4	29.64	217.15
GrO 5	28.05	217.09
GrO 6	30.3	217.03
GrO 7	29.62	220.28
GrO 8	31.49	218.76
GrO 9	31.86	214.87
GrO 10	30.00	217.12
Average	30.07	218.17
Std. dev.	1.16	1.82

Table 4.
TGA data of all GrO samples.

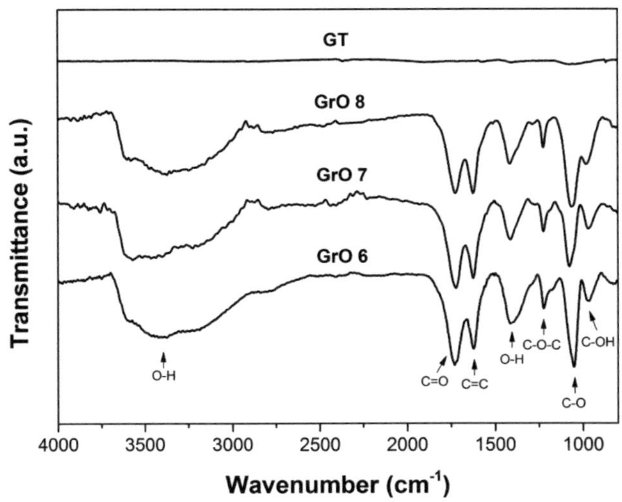

Figure 6.
FT-IR spectra of GT, GrO 6, GrO 7, and GrO 8 [17].

High-resolution XPS spectra of GrO in the C 1s region, shown in **Figure 8**, present the signals corresponding to nonoxygenated carbon rings (C═C/C—C, 284.7 eV), hydroxyl (C—OH, 286.3 eV), epoxy (C—O—C, 286.9 eV), carbonyl (C═O, 287.4 eV), and carboxyl groups (O═C—OH, 289.4 eV), which are consistent with the signals of the FT-IR spectra [16, 17, 23, 29].

Owing to the fact that several factors such as the nonstoichiometry nature of GrO, the size of the sheets, and no-homogeneity distribution of functional groups over the sheets and the fact that XPS is a technique of surface analysis, the concentration of oxygen functionalities among the specimens is not 100% reproducible, which has been consistent with previous reports; nevertheless, through other analysis techniques, it is demonstrated that the procedure for obtaining GrO is reproducible.

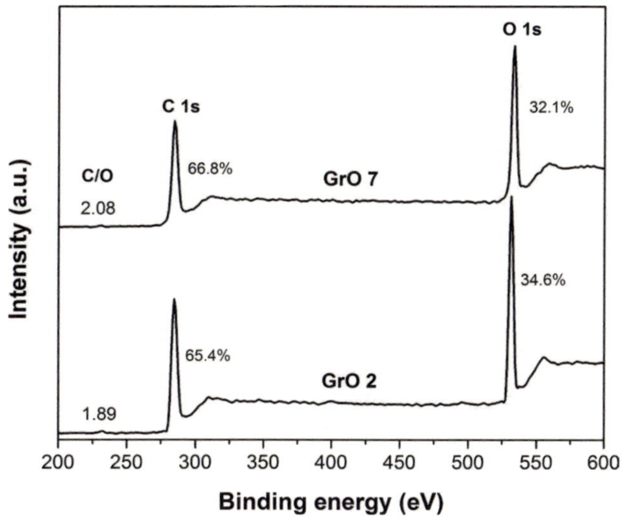

Figure 7.
XPS general survey of GrO 2 and GrO 7 [17].

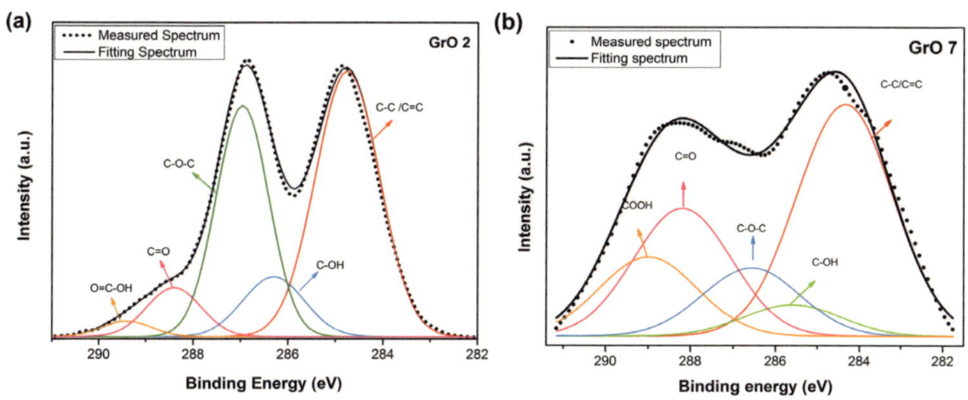

Figure 8.
HR-XPS survey spectrum of (a) GrO 2 and (b) GrO 7 [17].

3.4 Morphological properties

GrO powders were exfoliated in aqueous media via ultrasonic vibration, since it is one of the most common methods to exfoliate graphene oxide sheets [8], and the obtained samples were analyzed by TEM and AFM to monitor the morphology and the structure. The nanosheets and pure GT as well were studied by TEM in conventional mode (CTEM) and selected area electron diffraction mode (SAED), and **Figure 9a** shows a general view of GT, and it is clearly observed the plate-like shape, the borders are shown with different contrast, caused most likely for the presence of several plates randomly accommodated, which is not the case of graphene oxide, and these sheets tend to wrinkle and fold [30] and have a "silky" appearance. SAED pattern of GT (inset in **Figure 9a**) confirmed the polycrystalline nature of graphite, and the incident beam is surrounded for several rings with distinctive diffraction dots that correspond with some of its crystallographic planes, such as (201), (110), (100) and (101), according with the diffraction card: *PDF Card No.: 00-001-0646.*

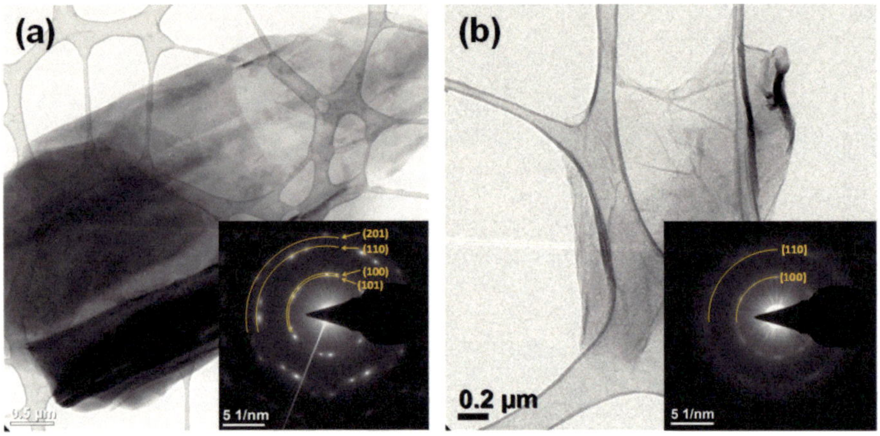

Figure 9.
TEM images of (a) GT and (b) GO, both with inset SAED showing the assigned planes [17].

The SAED pattern of GO (inset in **Figure 9b**) confirmed the disordered nature and shows a diffraction rings that are not well defined and unresolved dots, which is consistent with amorphous materials. Nevertheless, the measurements of these rings confirm the planes (100) and (110), which suggest the presence of regions with graphitic nature [23].

Atomic force microscopy (AFM) analysis was carried out to verify the number of layers of graphene oxide. The sample is collected from the dispersion prepared in deionized water, and this demonstrates that sonication promotes near-complete exfoliation of the GO; **Figure 10** exhibits an example of GO sheets with an estimated number of ~4 layers. Numerous nanosheets were detected in tapping mode, and the thickness profile showed around 1.45–6.42 nm, taking into consideration the *d*-space and the individual graphene sheet thickness (0.34 nm), leading to a conclusion that it was obtained GO with 1–6 layers, approximately [31].

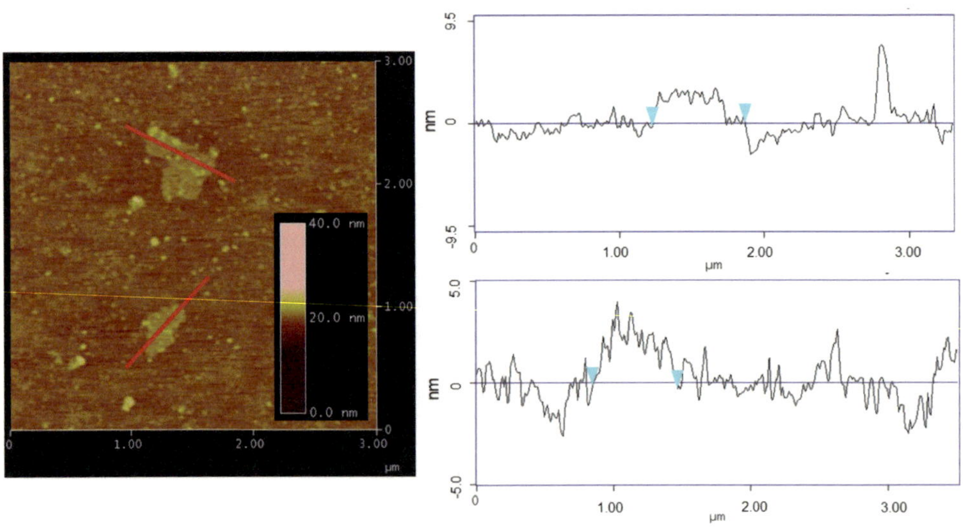

Figure 10.
Tapping mode AFM view of GO nanosheets with their thickness profile [17].

4. Conclusions

In summary, the protocol used to synthesize GrO based on Hummers' modified method proved to be a successful procedure. As has been demonstrated in early reports, the use of $NaNO_3$ is not a variable that influences the final product, and it is possible to achieve GrO with similar properties and characteristics by using a simple, small reaction time and more safe methodology. Besides the removal of $NaNO_3$, changes such as (a) enabling the sonication of graphite with sulfuric acid, (b) the slow addition of $KMnO_4$, and (c) the two-hour stirring of the mixture $KMnO_4$/graphite/sulfuric acid at 0°C were key factors that contributed to ensure reproducibility. This affirmation was supported by several analyses carried out to the 10 GrO samples synthesized. Oxygen functionality content, determined by TGA and XPS, was estimated of ~30 and ~33%, respectively; the type of these functional groups was identify by FTIR and corresponds to carboxyl, carbonyl, epoxy, and hydroxyl groups. Raman confirms that the layers of GrO have sp^2/sp^3 domains, suggesting that even though the oxidation has occurred, the sheets still possess graphenic characteristics; analyses performed by TEM pointed out that since SAED results showed graphenic ordering. This means that the material has the advantages of having functional groups in order to accomplish important chemical reactions (functionalization and reduction) while still possessing graphenic properties.

This proposed method can be used to synthesize GrO and other graphenic materials in an economical and large-scale way.

Acknowledgements

The support of the CONACYT through Grant 299,092 (LANIAUTO) is greatly appreciated. Also, the authors are grateful to Dr. Roberto Yañez Macías, Dr. Carlos Gallardo Vega, and Dr. Enrique Díaz Barriga Castro for their technical support.

Author details

Ernesto Hernández-Hernández*, Perla J. Hernández-Belmares, Mónica A. Ceniceros-Reyes, Oliverio S. Rodríguez-Fernández and Pablo González-Morones
Departamento de Materiales Avanzados, Centro de Investigación en Química Aplicada (CIQA), Saltillo, Coahuila, Mexico

*Address all correspondence to: ernesto.hernandez@ciqa.edu.mx

References

[1] Huang X, Yin Z, Wu S, et al. Graphene-based materials: Synthesis, characterization, properties, and applications. Small. 2011;**7**(14): 1876-1902. DOI: 10.1002/smll.201002009

[2] Graphene Market Size, Share & Trends Analysis Report by Application (Electronics, Composites, Energy), by Product (Graphene Nanoplatelets, Graphene Oxide), by Region, and Segment Forecasts, 2019-2025. 2019

[3] Elmarakbi A, Azoti WL. Novel composite materials for automotive applications: Concepts and challenges for energy-efficient and safe vehicles. In: 10th International Conference on Composites Science and Technology. 2015

[4] Lawal AT. Biosensors and bioelectronics Progress in utilisation of graphene for electrochemical biosensors. Biosensors & Bioelectronics. 2018;**106**(2018):149-178. DOI: 10.1016/j.bios.2018.01.030

[5] Sharma N, Sharma V, Jain Y, et al. Synthesis and characterization of graphene oxide (GO) and reduced graphene oxide (rGO) for gas sensing application. Macromolecular Symposia. 2017;**1**(376):1-5. DOI: 10.1002/masy.201700006

[6] Chen K, Wang Q, Niu Z, Chen J. Graphene-based materials for flexible energy storage devices. Journal of Energy Chemistry. 2018;**27**:12-24. DOI: 10.1016/j.jechem.2017.08.015

[7] Bai L, Zhang Y, Tong W, et al. Graphene for energy storage and conversion: Synthesis and interdisciplinary applications. Electrochemical Energy Reviews. 2019:1-36. DOI: 10.1007/s41918-019-00042-6

[8] Dreyer DR, Park S, Bielawski CW, Ruoff RS. The chemistry of graphene oxide. Chemical Society Reviews. 2010;**39**:228-240. DOI: 10.1039/b917103g

[9] Kuila T, Bose S, Kumar A, Khanra P, Kim NH, Joong Hee L. Chemical functionalization of graphene and its applications. Progress in Materials Science. 2012;**57**(7):1061-1105. DOI: 10.1016/j.pmatsci.2012.03.002

[10] Yu H, Zhang B, Bulin C, Li R, Xing R. High-efficient synthesis of graphene oxide based on improved Hummers method. Scientific Reports. 2016;**6**:36143. DOI: 10.1038/srep36143

[11] You X, Chang J-H, Ju BK, Pak JJ. An electrochemical route to graphene oxide. Journal of Nanoscience and Nanotechnology. 2011;**11**:5965-5968. DOI: 10.1166/jnn.2011.4451

[12] Pei S, Wei Q, Huang K, Cheng H, Ren W. Green synthesis of graphene oxide by seconds timescale water electrolytic oxidation. Nature Communications. 2018;**9**(145):1-9. DOI: 10.1038/s41467-017-02479-z

[13] Chen J, Li Y, Huang L, Li C, Shi G. High-yield preparation of graphene oxide from small graphite flakes via an improved Hummers method with a simple purification process. Carbon N Y. 2015;**81**:826-834. DOI: 10.1016/j.carbon.2014.10.033

[14] Hummers WJ, Offeman RE. Preparation of graphitic oxide. Journal of the American Chemical Society. 1958;**89**(9):1339. DOI: 10.1021/ja01539a017

[15] Marcano DC, Kosynkin DV, Berlin JM, et al. Improved synthesis of graphene oxide. ACS Nano.

2010;**4**(8):4806-4814. DOI: 10.1021/nn1006368

[16] Chen J, Yao B, Li C, Shi G. An improved Hummers method for eco-friendly synthesis of graphene oxide. Carbon N Y. 2013;**64**(1):225-229. DOI: 10.1016/j.carbon.2013.07.055

[17] Hernández BP, Rodriguez FO, Hernández HE. Elaboración de nanocompuestos de PA6/óxido de grafeno en el estado fundido, mediante el uso del concentrado PA6/rGO y de suspensions GO-H_2O [Tesis doctoral en proceso]. CIQA: Programa de Doctorado en Tecnología de Polímeros; 2016

[18] Lerf A, Buchsteiner A, Pieper J, et al. Hydration behavior and dynamics of water molecules in graphite oxide. Journal of Physics and Chemistry of Solids. 2006;**67**:1106-1110. DOI: 10.1016/j.jpcs.2006.01.031

[19] Buchsteiner A, Lerf A, Pieper J. Water dynamics in graphite oxide investigated with neutron scattering. The Journal of Physical Chemistry. B. 2006;**110**: 22328-22338

[20] Saenko NS. The X-ray diffraction study of three-dimensional disordered network of nanographites: Experiment and theory. Physics Procedia. 2012;**23**:102-105. DOI: 10.1016/j.phpro.2012.01.026

[21] Wang H, Hu YH. Effect of oxygen content on structures of graphite oxides. Industrial and Engineering Chemistry Research. 2011;**50**:6132-6137. DOI: 10.1021/ie102572q

[22] Yang D, Velamakanni A, Park S, et al. Chemical analysis of graphene oxide films after heat and chemical treatments by X-ray photoelectron and Micro-Raman spectroscopy. Carbon N Y. 2009;**47**:145-152. DOI: 10.1016/j.carbon.2008.09.045

[23] Krishnamoorthy K, Veerapandian M, Yun K, Kim SJ. The chemical and structural analysis of graphene oxide with different degrees of oxidation. Carbon N Y. 2013;**53**:38-49. DOI: 10.1016/j.carbon.2012.10.013

[24] Pimenta MA, Dresselhaus G, Dresselhaus MS, Cancado LG, Jorio A, Saito R. Studying disorder in graphite-based systems by Raman spectroscopy. Physical Chemistry Chemical Physics. 2007;**9**:1276-1291. DOI: 10.1039/b613962k

[25] Shen J, Hu Y, Shi M, et al. Fast and facile preparation of graphene oxide and reduced graphene oxide nanoplatelets. Chemistry of Materials. 2009;**21**(15):3514-3520. DOI: 10.1021/cm901247t

[26] Chu H, Lee C, Tai N. Green reduction of graphene oxide by *Hibiscus sabdariffa* L. to fabricate flexible graphene electrode. Carbon N Y. 2014;**80**:725-733. DOI: 10.1016/j.carbon.2014.09.019

[27] Acik M, Lee G, Mattevi C, et al. The role of oxygen during thermal reduction of graphene oxide studied by infrared absorption spectroscopy. Journal of Physical Chemistry C. 2011;**115**: 19761-19781. DOI: 10.1021/jp2052618

[28] Rattana T, Chaiyakun S, Witit-Anun N, et al. Preparation and characterization of graphene oxide nanosheets. Procedia Engineering. 2012;**32**:759-764. DOI: 10.1016/j.proeng.2012.02.009

[29] Yuan R, Yuan J, Wu Y, Chen L, Zhou H, Chen J. Efficient synthesis of graphene oxide and the mechanisms of oxidation and exfoliation. Applied Surface Science. 2017;**416**:868-877. DOI: 10.1016/j.apsusc.2017.04.181

[30] Pandey DK, Fung T, Prakash G, Piner R, Chen YP, Reifenberger R. Surface science folding and cracking of

graphene oxide sheets upon deposition. Surface Science. 2011;**605**:1669-1675. DOI: 10.1016/j.susc.2011.04.034

[31] Frankberg EJ, George L, Efimov A, Honkanen M, Pessi J, Levänen E. Measuring synthesis yield in graphene oxide synthesis by modified Hummers method. Fullerenes, Nanotubes, and Carbon Nanostructures. 2015;**23**(9):755-759. DOI: 10.1080/1536383X.2014.993754

Chapter 4

Stabilized Graphene Oxide Assisted Surfactants and Its Capacitance Performance

Nurhafizah Md Disa

Abstract

The use of surfactant in achieving high stabilization and exfoliation graphene oxide (GO) was seen to be the crucial factor in improving the quality and quantity of GO produced via electrochemical exfoliation method. Therefore, this chapter is presenting the physical characterizations that successfully showed the stabilization of GO by various anionic surfactant. The uniqueness in this work was lie on the number of tails surfactant which consists of single-, double- and triple-tails, and the comparison study in terms of electrical conductivity between the commercial and custom-made surfactants. Moreover, the effective stabilized-GO was further explored in super capacitor application. This work opens a new window for green and low-cost GO material as conductive electrodes material.

Keywords: graphene oxide, surfactant, electrical conductivity, supercapacitor

1. Introduction

In 2004, Graphene (GE) has evolved as an interesting two-dimensional gapless semiconductor material [1]. GE is suitable for many technological applications due to the fascinating properties such as high mobility of charge carriers, large surface area calculated up to 2630 m^2g^{-1}, high mechanical strength (young modulus ~ 1 Tpa), excellent thermal conductivity and almost transparent [2–5]. However, the issue of graphene's price due to the world energy crisis intensively affects the national progress of developing countries, because it weighed down all sectors, such as in economy, industry, agriculture, social life and inflation to poverty, hence national progress slow down. The demand of GE has gained attention even though the price is tremendously increased year by year and the price predicted to exceed USD 200 million by 2024 according to Global Market Insight, Inc. The increasing factors are due to the demands from various industries including automotive, aerospace and electronics. Thus, the implementation of its derivative from the graphene's family is highly needed to cater the economy issue. Thus, in recent years, Graphene oxide (GO), and reduced GO (rGO) have gained much attention due to its superb properties. The existence of oxygen content that is bonded on the surfaces and edges part of GO sheets is important for functionalization purpose. So far, various techniques were carried out to produce GO from various carbon sources such as scotch tape, CVD, chemical reduction process as well as carbonization of natural sources. Because of that, the development of methods that allow the mass production of excellent quality graphene materials has become a top priority today. The GO material is now emerging from the laboratory into the commercial product, soon our nation will be recognized as a key player involved in this frontier technology. However, the prevention of aggregation was of particular importance for GE sheets because most of their unique properties were only associated with individual sheets and keeping them well separated was required. Researchers have found that the maximum improvement of GE properties can totally be seen by compositing GE along with other conducting materials such as carbon nanotubes (CNTs) in a small quantity [6].

To date, GE dispersions in the polymer matrix performed higher electrical properties which beyond the mixing of two conventional nanofillers namely silicates and carbon nanotubes (CNTs) [7, 8]. However, the highly restacking GE-based materials due to the strong van der Waals interactions between the GE sheets, led this kind of materials to have low solubility in the polymer matrix. This paved the way to introduce such interfacial adhesion in order to produce finely dispersed GE-based materials in polymer matrices and believed to improve the nanocomposite conductivity [9–12].

Therefore, the use of non-functionalization of surfactant to assist GE-based materials dispersions in the nanocomposite was currently employed and reported to highly preserve the properties of nanofillers in the nanocomposite as compared to covalent modifications [13, 14]. In addition, the existence of the excess surfactant in the nanocomposite was highly affected the electrical conductivity of the end-product of nanocomposite even though some researchers reported that the surfactant hinder the electron movement of GE in the nanocomposite [15]. Apart from that, the number of tail group of surfactant was also seen to have an important impact to produce better dispersion subsequently, leading to increased electrical conductivity [14]. They prepared the thermally reduced GO (TRGO) using annealing process at temperature of about ~1000°C which further stabilized by ionic sodium dodecyl sulfate (SDS) and non-ionic (Pluronic F 127) surfactants. The nanocomposite was obtained by means of latex technology and showed higher electrical percolation threshold of around 10^{-6} S cm^{-1} (SDS) as compared to Pluronic F127 stabilized in TRGO/natural rubber latex (NRL) polymer nanocomposite (~10^{-9} S cm^{-1}). This results further gives information that the ionic SDS surfactant is an effective surfactant in stabilizing GE-based materials led in good dispersion of nanocomposite produced.

In addition, Mohamed et al. [16, 17] also demonstrated the electrical improvements of the CNTs/NRL polymer nanocomposite by increasing the number of tail group of surfactant from single- (SDS) to double- (AOT4) and triple-chain (TC14). They found that electrical enhancement was increased up to ~× 10^{-3} S cm^{-1} for the CNTs/NRL polymer nanocomposite assisted TC14 as compared to pure NRL (σ = ~10^{-11} S cm^{-1}). Due to surfactant stabilization theory of CNTs-based nanocomposite was similar to the GE-based nanocomposite, the probability similar effect of GE-based materials dispersion filled in the NRL polymer with the assisted various number of tail groups might also led to the significant improvement in the electrical conductivity and capacitance performance.

Until now, the lack of studies regarding to the properties and intermolecular interactions of GE assisted various number of surfactant tail groups in the nanocomposite has made this field loosely explored. Moreover, GE-based materials always show re-stacking properties when forming an agglomeration structures. Due to the inaccessible site for ions diffusion in restacking agglomerate GO, the use of so-called non-functionalization of surfactants were demonstrated to encounter the excess agglomerated sheets in the nanocomposite electrode materials. With the addition of surfactants for stabilizing GO or rGO sheets during fabrication process in the nanocomposite, researchers have found the effective SDBS as stabilized agent in the nanocomposite which performed the highest stability using *C-V* measurement [18]. Yuan et al. also demonstrated the higher specific capacitance of 690 F g^{-1} in the presence of SDBS-stabilized GE with PANI as electrodes materials in supercapacitor [19]. In addition, Jothi et al. obtained 1312 F g^{-1} of capacitance performance by applying pluoronic P123 in the hybridization between rGO and nanoporous nickel sulfide [20]. They reported three main factors which contribute to the high capacitance performance, including: (i) high porous structure of material, (ii) uniform surface coating and (iii) reduced stacking effects of rGO sheets.

Thus, it is crucial to properly select the interfacial interaction molecule to assist the dispersion of GE-based materials. The surfactant has stabilized GO sheets more in the nanocomposite and thus can increase the effective platform for ion kinetics movements between the electrodes and the electrolyte. Moreover, the tail group at the hydrophobic part of the surfactant was also responsible for the specific capacitance increment. The increasing number of tail groups might believe to help and assist carbon group more effective in the NRL polymer. Thus, a more convenient, green, and simple approach to fabricate nanocomposite with similar or beyond the electrical conductivity reported and improved capacitive behavior needs to be seriously investigated.

Therefore, in this chapter, the aim of the work is to further discuss on the effects of different type of surfactants in GO production. The work was systematically performed and discussed accordingly throughout the chapter. The synthesize of GO were carried out using commercially available surfactant of single-tail (SDS, SDBS, PSS) and custom-made surfactants of double- (AOT4) and triple-tails (TC14) at 7 V (0.1 M) for 24 hours synthesis process via electrochemical exfoliation method [21, 22]. Meanwhile, the nanocomposite samples were fabricated using one-step approach of electrochemical exfoliation as reported previously [23, 24]. The performance of all samples was discussed in terms of its electrical and capacitive behavior.

2. Graphene oxide assisted single-, double-, and triple-tails of surfactants

In this part, the characterizations of GO production using various surfactants were demonstrated; including morphological surfaces (FESEM), quality and defects level (Raman), absorption properties (UV-Vis). This stage is very important for quality and quantity evaluation of synthesized GO before incorporated with NRL polymer latex. All samples were measured using field emission scanning electron microscopy (FESEM- Hitachi SU 8020), micro-Raman spectroscopy (Renishaw InVia Raman Microscope), UV-Vis spectroscopy (Agilent 8453 Spectrophotometer).

2.1 Morphological characterization

FESEM is the most important tool to figure out the morphological characteristics of GO produced. From the FESEM images, the GO samples synthesized using 0.1 M of various surfactants are presented in **Figure 1(a)–(e)** The commercially available single-tail surfactants of SDS, SDBS and PSS were the initial type of surfactant used for electrolyte preparation. Generally, these three single-tail surfactants gave a crumple layers of GO. Between those three single-tail surfactant (SDS, SDBS and PSS), the growth of well-structured GO was observed at sample prepared using SDS surfactant (**Figure 1(a)**). This was predicted due to the low-molecular surfactants that pack tightly on the GO surfaces for maximizing the surface charge

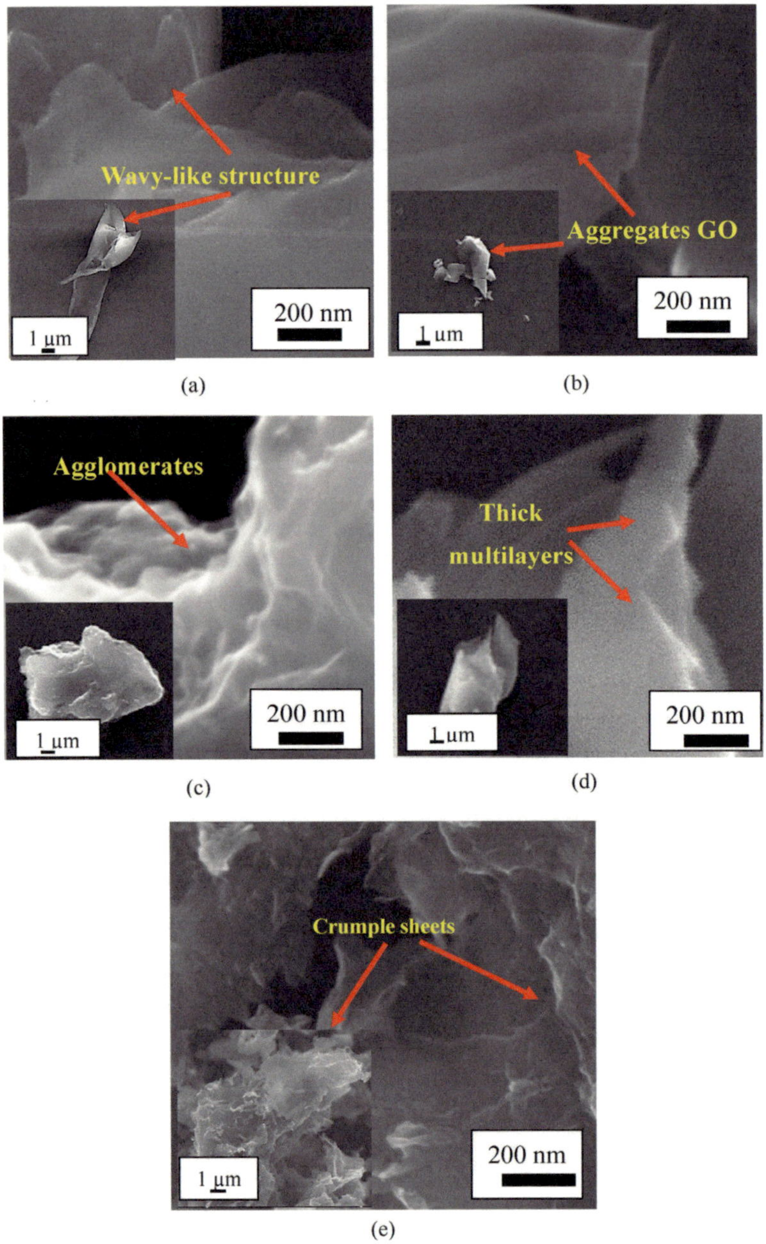

Figure 1.
FESEM images of 0.1 M-GO synthesized using different surfactants (a) SDS, (b) SDBS, (c) PSS, (d) AOT4, and (e) TC14 via electrochemical exfoliation method.

in the graphitic layers. This has led to the optimum exfoliated GO that can be achieved to be in the dispersions [25, 26].

Meanwhile, the SDBS-GO sample shows the fold up structure at the edges of GO sheets and the sample has the tendency to aggregate as compared to the SDS (shown in inset **Figure 1(b)**). This was probably because the benzene ring in the SDBS molecule causes a π-π interaction between molecules in addition to hydrophobic interaction between surfactant tails [27]. Therefore, at higher concentration of electrolyte, the number of the interactions between the surfactant and the GO sheets was increased which causes to the formation of aggregation GO sheets and stacked layers at the basal plane.

In addition, the PSS-GO sample (**Figure 1(c)**) also shows agglomerated structure of GO. This observation was due to the relatively weaker π-π interactions between the surfactant molecule and the GO surfaces thus did not effectively reduce the agglomeration of GO structures [28]. This shows that the benzene ring located in the head group of surfactant does not give significant effects to the GO produced due to the fact that the hydrophobic tails group was mainly responsible for the intercalation and interaction between the GO sheets and the tail group of the surfactant. This result was consistent with the fact that at higher electrolyte concentration which represents the high amount of surfactant, largely led to the agglomerations of GO sheets. Therefore, the use of single-tail surfactant (SDS, SDBS and PSS) at higher concentration only shows slightly improve in the GO dispersions with slightly higher GO volume produced. Due to the important role of the hydrophobic tails group surfactant in the exfoliation process of GO, the custom-made surfactant of double- and triple-tails of AOT4 and TC14, respectively was further investigated.

In comparison to the double- and triple-tails surfactant of AOT4 and TC14 as shown in **Figure 1(d** and **e)**, the TC14-GO sample produces at 0.1 M shows more crumpled structure as compared to the double-tails surfactant of AOT4-GO sample (**Figure 1(d)**). This might believe due to the triple-tails surfactant offers triple interaction and stabilization on the GO surfaces which led to the maximum exfoliated GO in the sample produced as compared to the single- (SDS, SDBS and PSS) and double-tails (AOT4) surfactant. At approximate volume of 0.1 M, the triple-tails of TC14 inserted helps to triple order separate the graphitic layers and prevent GO layers from reforming the π-π stacking interaction between the GO sheets. The lower surface tension of triple-tails TC14 (γ_{cmc} = 27.0 mN m^{-1}) attempting to minimize the hydrophobic interactions between GO layers which subsequently find the way to the gaps opening up at the edges more than the higher surface tension of single-tail SDS (γ_{cmc} = 34.7 mN m^{-1}). Due to the different surface tension of different surfactants, the single-tail surfactants were seen not sufficient to dislodge the graphite layers to individual sheets and bring the GO sheets into the water-based medium. Meanwhile, for double-tails surfactant, AOT4-GO sample shows a smooth GO surface exfoliated to the multilayers as pointed by arrows. However, the thick layers observed have convinced that the double-tails surfactant still beneath the sufficient level to successfully exfoliate the GO sheets to thin and transparent layers. Therefore, the length and greater branches of alkyl chains of surfactant play a role in the GO dispersions [28–30].

Micro-Raman spectroscopy has a sensitivity, and it is the most general technique to characterize the carbonaceous materials. This technique was important in order to observe the vibrational and rotational of the carbon materials. As previously discussed, it was well-known that higher crystallinity of the GO produced was shown by the higher intensity ratio of D- to G-peaks due to successful oxidation process [31]. On the other hand, the increased of the I_D/I_G ratio can also be measured as the increase of the defect levels of the sample. This can be reflected to the PSS-GO sample whereas the lowest I_D/I_G ratio of 0.64 and sharp 2D-peak at 2712.3 cm^{-1} has indicated the GO sample can also produce a better crystallinity at higher concentration. The obviously intense shape of the 2D-peak in PSS-GO dispersions shows that the dispersions were also consisted by several single GE layers [32, 33]. However, the PSS-GO sample has shown the additional band (G+ band) which detected at ~1620 cm^{-1} represented the edge defects. This additional peak was corresponded to the edges of GO sheets fold up due to the π-π interaction between the aromatic rings of the surfactant and the GO sheets. This resulted from the low energy of the surfactant used to exfoliate GO at the edges part by using higher electrolyte concentration (0.1 M). The repetition on micro-Raman measurement of PSS-GO sample shows that the calculated I_D/I_G ratio value were not consistent across the samples suggesting low structural uniformity. The observation has been supported by the FESEM where the sample has formed an agglomerates GO sheets and less uniformity. Due to this, the single-tail surfactant of PSS was not included in the next GO growth parameter.

Meanwhile, for SDBS-GO sample, a faint G+ band was observed at around ~1639 cm^{-1} in the SDBS-GO sample. This shows that the single-tail surfactants of SDBS and PSS were not enough to exfoliate the GO sheets into the single and thin layers of GO. Meanwhile, by increasing the number at the tail group of surfactant from double- (AOT4) to triple-tails (TC14), the decrease pattern of I_D/I_G ratio was observed which were calculated to be 0.88 and 0.82, respectively. The moderate quantity and quality of the TC14-GO sample indicated that the triple-tails surfactant shows a great improvement in the exfoliating and stabilizing GO sheets. This might be explained by the triple interactions of surfactant to assist the production of GO sheets via electrochemical exfoliation method.

In addition, less intense 2D-peak appeared in the TC14-GO sample at around 2715.8 cm^{-1} shows that the GO produced also consists of few layer GO sheets. However, the biggest 2D-peak width (96.3 cm^{-1}) indicated that the GO sheets were highly stabilized by the triple-tails TC14 surfactant thus increased the interlayer spacing of GO layers. In comparison, upon the introduction of single- and double-tails surfactant of SDS and AOT4 assisted in the GO production, a broad and weak 2D-peak (~2712.8–2718.0 cm^{-1}) were detected which indicates that the GO produced has a slightly higher of structural disorder of GO [35]. The details have been shown in **Figure 2** and **Table 1**. The results were consistent with the FESEM analysis as discussed above where the number of tail group and the electrolyte concentration of surfactant plays an important role in the high yield and better GO sample production and stabilization.

UV-Vis absorption measurement was conducted on the synthesized GO and is shown in **Figure 3**. The absorption peaks at around ~223–232 nm in the UV-Vis spectrum was detected in the synthesized GO using single- (SDS, SDBS and PSS),

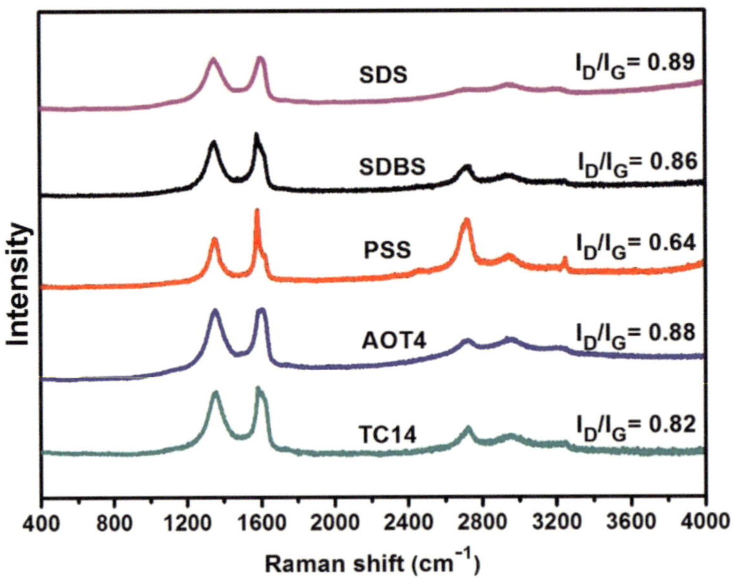

Figure 2.
Micro-Raman analyses of GO production using 0.1 M at various types of surfactant (SDS, SDBS, PSS, AOT4, and TC14) via electrochemical exfoliation method [34].

Samples: Different surfactants	G-peak (cm^{-1})	G-peak width (cm^{-1})	D-peak (cm^{-1})	D-peak width (cm^{-1})	I_D/I_G ratio	2D-peak	2D-peak width (cm^{-1})
0.1 M SDS-GO	1598.2	58.5	1357.0	97.7	0.89	2712.8	59.3
0.1 M SDBS-GO	1590.7	61.4	1353.3	91.2	0.86	2716.6	57.7
0.1 M PSS-GO	1582.3	26.9	1353.2	48.4	0.64	2712.3	73.2
0.1 M AOT4-GO	1597.6	59.4	1357.6	89.9	0.88	2718.0	69.7
0.1 M TC14-GO	1595.5	65.8	1358.3	120.2	0.82	2715.8	96.3

Table 1.
Micro-Raman analyses for GO produced using 0.1 M at various types of surfactant (SDS, SDBS, PSS, AOT4, and TC14) via electrochemical exfoliation method.

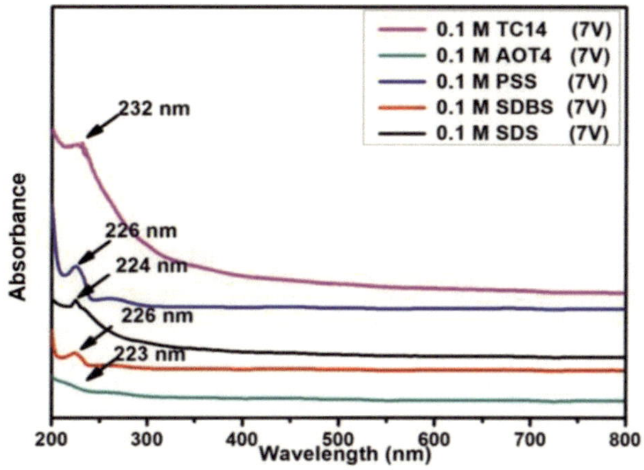

Figure 3.
UV-Vis spectra of GO produced using 0.1 M for different type of surfactants (SDS, SDBS, PSS, AOT4, and TC14) via electrochemical exfoliation method.

double- (AOT4) and triple-tails (TC14) surfactant. In comparison to the lower electrolyte concentration of 0.01 M, overall samples show the shifted peak to a high wavelength except for the sample prepared using SDS and AOT4 surfactant. It can be noted that at higher electrolyte concentration, the amount of GO produced has increased thus increasing the formation of various oxygen functional groups as confirmed by the slightly shifted peak to higher wavelength in UV-Vis.

Meanwhile, the prepared TC14-GO sample shows the obvious shifted peak of GO from 220 nm to highest wavelength of 232 nm as compared with other GO samples which attributes a high level of stability achieved by the GO dispersions [36]. On the other hand, the shifted peak at higher wavelength also as indicator of good dispersion of GO sheet where the largest amount of exfoliated GO was suspended in the water [35]. This characteristic shows that the triple-tails surfactant TC14 maintained the high ability to stabilize the GO sheets at both low (0.01 M) and high (0.1 M) electrolyte concentration. By increasing the electrolyte concentration from 0.01 to 0.1 M, the synthesized TC14-GO sample possesses the highest consistency in term of stabilization of GO sheets. In addition, it was believed that the nanoplatelets GE were also produced instead of sheets structures in the sample [36].

This information might be beneficial to the increase the interactions thus leading to the enhancement in electrical conductivity and capacitance performance.

In conclusion, GO synthesized using 0.1 M of triple-chain TC14 surfactant was successfully produced a highly crumpled structure with moderate crystallinity (I_D/I_G = 0.82) as compared to the single-chain SDS surfactant where a slightly higher crystallinity (I_D/I_G = 0.89). The triple-chain TC14 surfactant probably has well-stabilized the exfoliated GO sheets as compared to the single-chain SDS during the electrochemical exfoliation method. The triple-tails of TC14 might give the triple interactions to the GO production thus increasing the stabilization of GO sheets. These results have been confirmed by the UV-Vis analysis by the shifted peak from 220 to 232 nm. Therefore, GO-assisted triple-tails TC14 surfactant was considered as the suitable surfactant for higher quality and reasonably good graphitized GO as compared to single- (SDS) and double-tails (AOT4) surfactant.

3. Graphene oxide/natural rubber latex polymer nanocomposite synthesized via one-step method using different type surfactants

The one-step approach was implemented due to a simpler and rapid production of GE-based electrodes. Through the one-step, upon the production of GO solution, latex milk was simultaneously mixed in the exfoliation system for overnight. The obtained sample was dried for 24 hours in a petri dish to form a thin film. This work has been performed as previously discussed [37, 38]. The electrical conductivity and capacitive behavior was done using four-point probe (Keitley 2636A), and cyclic voltammetry (CV-Gamry potentiostat series-G750 equipped galvanostat) measurements, respectively. Moreover, the morphological characterization using FESEM was also discussed in this section in order to support the stability information of the electrodes produced.

3.1 Electrical enhancement

Figure 4 depicts the electrical conductivities of the TC14-GO/NRL and SDS-GO/NRL polymer nanocomposite. The *I-V* curves presented in **Figure 4** revealed that increasing the number of single-tail SDS to triple-tail TC14 surfactant led to significant enhancement of electrical conductivity. This finding was consistent with

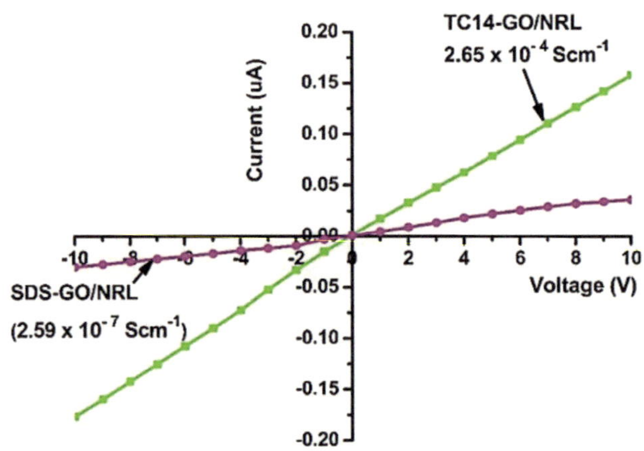

Figure 4.
I-V curves of the 0.1 M GO/NRL polymer nanocomposite stabilized using SDS and TC14 surfactant synthesized via one-step method of electrochemical exfoliation.

that of Mohamed et al. in dispersing CNT in the NRL matrix by the latex technology approach [29]. The presence of TC14 surfactant resulted in the high electrical conductivity range (2.65×10^{-4} S cm^{-1}) compared with the single-chain SDS surfactant (2.59×10^{-7} S cm^{-1}). This result showed that the triple-chain TC14 surfactant assisted in the formation of conductive pathways between the GO and NRL matrix [39]. The conductive network was formed by allowing the hydrophobic part of the surfactant to adsorb and interact with the GO structures, thereby decreasing the surface charge of the carbon material [40]. Therefore, the electrical conductivity improvement of the GO/NRL polymer nanocomposite was strongly affected by the tail group of the surfactant, which was believed to improve the dispersions between GO and NRL polymer. This result suggests that the nature of surfactant plays an important role to obtain GO/NRL polymer nanocomposite with significantly higher electrical conductivities. This finding was also supported by the capacitance behavior where the TC14-GO/NRL polymer nanocomposite samples showed high current response and large leaf shape.

3.2 Capacitance performance

Figure 5**(a)** and **(b)** shows the *C-V* curves of TC14-GO/NRL polymer nanocomposite prepared at a scan rate of 100 m Vs^{-1}. A high capacitance value of 25 F g^{-1} was measured for the TC14-GO/NRL polymer nanocomposite sample as compared to the SDS-GO/NRL polymer nanocomposite sample (5 F g^{-1}). The triple-tails surfactant, TC14 shows the higher influence on the capacitance behavior of GO/NRL polymer nanocomposite as compared to single-tail surfactant, SDS. The resulted capacitance obtained was due from good ion propagation between the TC14-GO/NRL polymer nanocomposite electrodes and the electrolyte. This might be explained as the single-tail SDS surfactant has low interaction with ratio 1:1 between the GO sheets and the NRL polymer matrix as compared to the triple-tails TC14 surfactant where triple-tails surfactant has triple interactions to GO sheets with ratio 3:1. In addition, the low capacitive performance obtained shows that at higher concentration of electrolyte (0.1 M), more agglomerations formed were led to the higher resistivity of the ion transfer between the electrolyte and electrodes. Therefore, the triple-tails TC14 surfactant has built triple pathways for ions diffusion between the electrodes and the electrolyte. The remained surfactant in the nanocomposite has efficiently prevented aggregation and restacking of GO upon cyclic measurement. Furthermore, the single-tail SDS surfactant also has low ion mobility in aqueous solution which limited the charge diffusion between the electrodes and the electrolyte [41]. The type of surfactant used has great influenced on the ions mobility of medium within the electrolyte.

However, the *C-V* curves of both GO/NRL polymer nanocomposite shows slightly distortion of the leaf-like shape, indicating that the surfactants introduced have positive effect on capacitance behavior of the nanocomposite electrode. It also can be seen from **Figure 5** that with the addition of triple-tails surfactant, TC14, the *C-V* curve becomes broader which subsequently improved the capacitance behavior. These results indicated that the TC14-GO/NRL polymer nanocomposite produced have great potential to be implemented as green and low-cost supercapacitor electrodes.

In the CD curves, the longer discharge time than the charging time at low current density implies an asymmetric behavior in charging and discharging section, due to the redistribution of charges after charging the electrode [42, 43]. From the results (**Figure 6**), the charging and discharging time pattern showed a symmetric behavior, which concluded that both electrodes have potential in restoring the charges. However, in comparison, the sample consisted triple-tails,

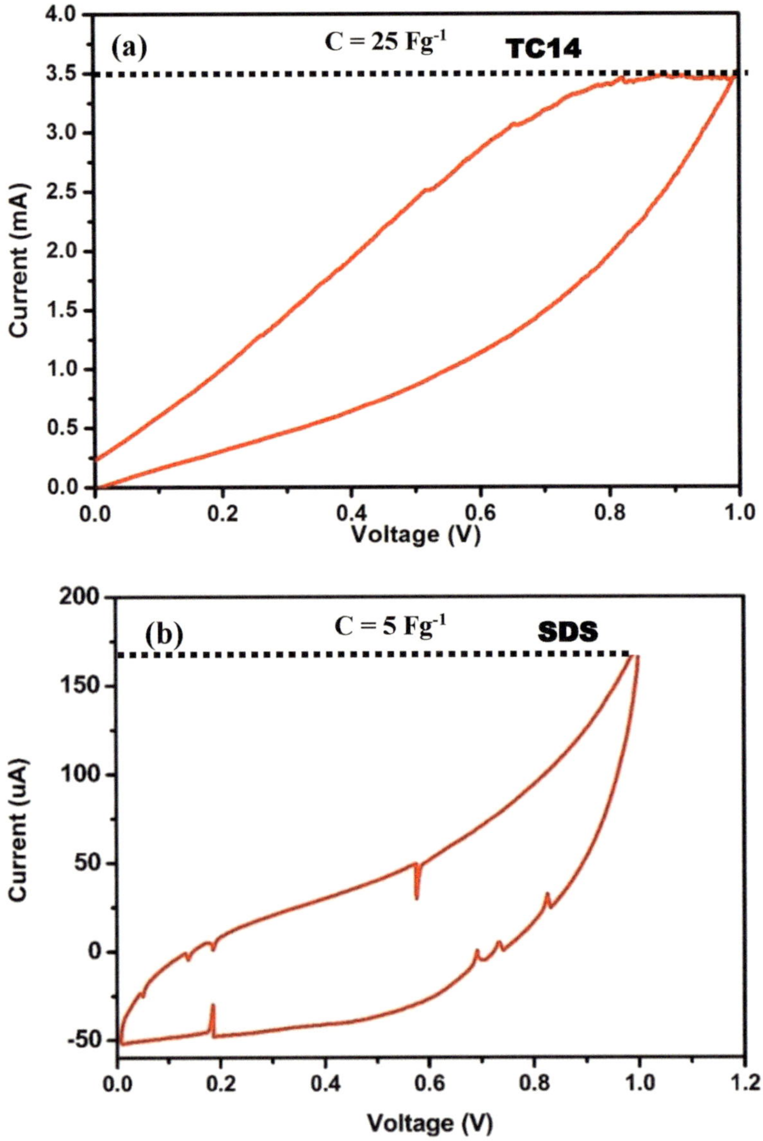

Figure 5.
C-V curves of the (a) TC14-GO/NRL and (b) SDS-GO/NRL polymer nanocomposite synthesized via one-step method of electrochemical exfoliation.

TC14 (**Figure 6(a)**) showed a distortion pattern between charging and discharging time which probably due to the high surface redox reaction of oxygen-containing functional groups [44, 45]. This was supported by the specific capacitance obtained by both nanocomposites. Meanwhile, the low specific capacitance observed by the sample of 0.1 M SDS-GO/NRL was expected due to the jumbled distribution of GO samples on the NRL matrix surface which was supported by FESEM image. This has led to the increment of resistance of both charge transport and electrolyte ions diffusion [46]. The crack formation between the GO sheets and NRL polymer after cycling process also convinced that the nanocomposite has low interaction. However, in general, the discontinuous conductive network between the GO sheets assisted surfactant in the NRL matrix also bring to the low of specific capacitance and charge-discharge pattern.

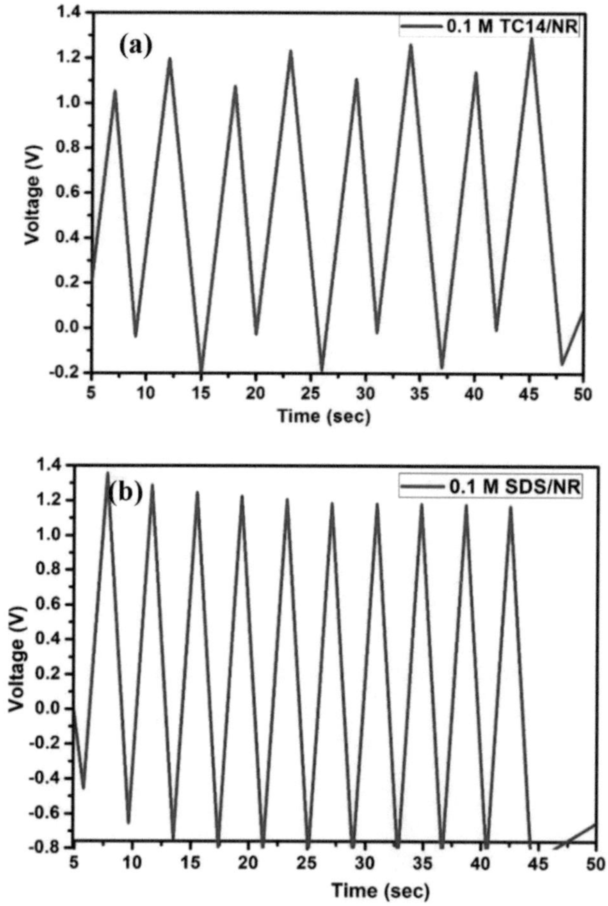

Figure 6.
C-D curves of the (a) TC14-GO/NRL and (b) SDS-GO/NRL polymer nanocomposite synthesized via one-step method of electrochemical exfoliation.

3.3 Morphology characterization

Figure 7 shows the cross-sectional FESEM images of the SDS-GO/NRL and TC14-GO/NRL polymer nanocomposite samples were studied to investigate the dispersion of the GO sheets assisted surfactant into the NRL polymer matrix. The figure clearly demonstrations that both the nanocomposite samples show dispersion of the few layers of exfoliated GO sheets throughout the NRL matrix. Dispersion GO assisted triple-tails surfactant, TC14 plays a significant role to fabricate high performance GO/NRL polymer nanocomposite. As reported in previous work [47], a good dispersion in the polymer matrix can be achieved by including surfactant in the composite materials. This was due to the improvement in the hydrophobic interaction between the surfactant and GO sheets layer, thus lowering the van der Waals between the graphene sheets. After sometimes of cycling testing, a crack surface in SDS-GO/NRL polymer nanocomposites sample observed in between the agglomerated GO sheets and polymer matrix (**Figure 7a**). This might contribute to the decreasing patent of charging-discharging measurement (C-D curve in **Figure 6**). In comparison, the higher dispersion achieved by the TC14-GO/NRL nanocomposites and higher stability in C-D curve showed that a great influence of high branches surfactant in increasing the long-time stability of the nanocomposite produced. A bulk GO sheets observed confirmed a higher interaction of triple-tails TC14 to the GO surfaces, thus consequence to the bulk agglomerated GO.

Figure 8 gives the illustration of the GO sheets produced using single-tail surfactant of SDS at both low (0.01 M) and high (0.1 M) electrolyte concentration.

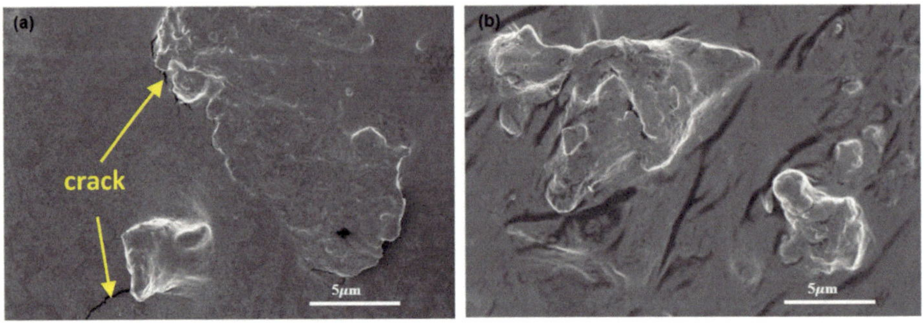

Figure 7.
FESEM images of the (a) SDS-GO/NRL and (b) TC14-GO/NRL polymer nanocomposite synthesized via one-step method of electrochemical exfoliation. Noted that the GO sheets are shown by the pull out bulk particle from the smooth NRL matrix.

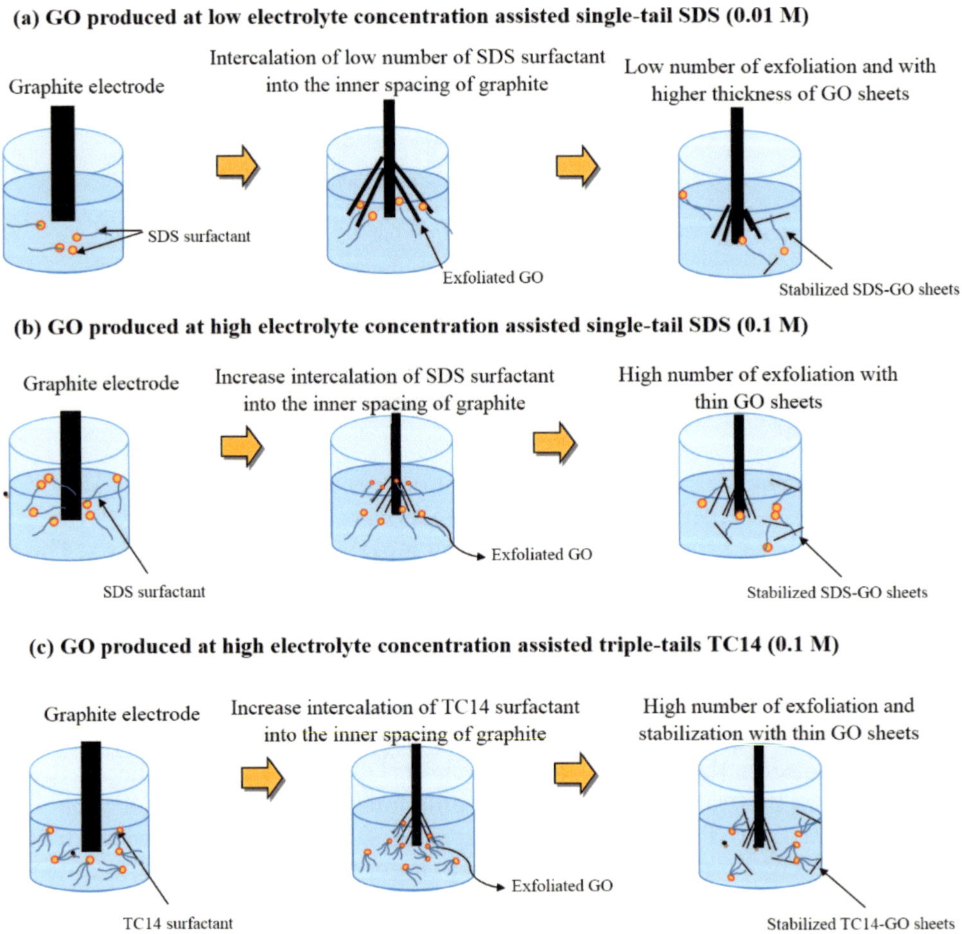

Figure 8.
Illustration of (a) 0.01 M SDS-GO, (b) 0.1 M SDS-GO, and (c) 0.1 M TC14-GO samples produced via electrochemical exfoliation method.

The schematic diagram in **Figure 8(a)** shows that the SDS surfactant has exfoliated the graphite into several layers of GO. At low electrolyte concentration of 0.01 M SDS, the low number of SDS surfactant yielded low amount of GO produced. The low number of surfactant has caused to the insufficient energy to successfully exfoliate the GO layers into thin sheets thus produced an agglomerate and thick GO sheets. As compared to high electrolyte concentration of 0.1 M (**Figure 8 (b)**), the higher number of SDS surfactant has increased the amount of exfoliated GO sheets in the dispersion. This was due to the sufficient energy of SDS at above critical micelle concentration (CMC) level to exfoliate graphite into multilayers of GO and the SDS has further stabilized the layers as shown in **Figure 8(b)** Due to the strong van der Waals interaction between the GO layers, the highly exfoliated layers of GO has tends to stack on top of the GO layers thus producing thicker GO sheets with numerous wrinkles. This illustration was consistent with the FESEM images where the produced GO at high concentration (0.1 M) has slightly bulk layers indicates high number of exfoliated GO. Meanwhile, **Figure 8(c)** compares the greater branches of triple-tails surfactant, TC14 assisted in the GO production using electrolyte concentration of 0.1 M with the commercially available single-tail, SDS surfactant (0.1 M). At above of CMC level of TC14 surfactant (0.1 M), the TC14 surfactant provides triple order interaction and higher stabilization between the triple-tails surfactant with the GO sheets thus produced higher volume and great quality of GO sheets compared to SDS-GO. The use of triple-tails surfactant of TC14 at and above the CMC level may facilitates the similar triple interaction to the GO sheets in both aqueous and non-aqueous electrolytes where the TC14 has shown important role in determining the number of thin GO sheets with higher stabilization of the GO layers produced.

4. Conclusion

Various analytical techniques done revealed that the GO produced were strongly affected by the surfactants used (TC14, AOT4, SDBS, PSS and SDS). TC14 surfactant was considered the most suitable surfactant for higher quantity and reasonably good graphitized GO (I_D/I_G = 0.82–0.83). Due to the high quantity and quality of GO produced using triple-tails TC14 surfactant, a close comparison was done with the single-tail SDS surfactant in order to investigate the performance of GO-assisted surfactant in the NRL polymer nanocomposite. The one-step method with 24 hours of synthesis time was carried out at 7 and 10 V using 0.1 and 0.01 M, respectively, throughout this work. The initial analysis was done by measuring the electrical properties of GO filled NRL nanocomposites then followed by capacitance and structural properties. This was due to the samples were bench making by its electrical conductivity then only capacitance and structural properties.

Acknowledgements

The author wants to acknowledge Short Term Grant, Universiti Sains Malaysia (USM, Penang) (304/PFIZIK/6315241) that has made this research possible. Special thanks to Nano-Optoelectronics Research and Technology Laboratory, School of Physics (USM) and Nanotechnology Research Centre, Universiti Pendidikan Sultan Idris (UPSI) for their facilities support. Author is thankful to Malaysian Rubber Board industry for latex supplied and many thanks to Associate Professor Dr. Azmi Mohamed for the custom-made surfactants supplied in this study.

Conflict of interest

The authors declare no conflict of interest.

Author details

Nurhafizah Md Disa
School of Physics, Universiti Sains Malaysia, USM Penang, Malaysia

*Address all correspondence to: mdnurhafizah@usm.my

References

[1] Novoselov KS, Geim AK, Morozov SV, Jiang D, Katsnelson MI, Grigorieva I, et al. Two-dimensional gas of massless dirac fermions in graphene. Nature. 2005;**438**:197-200. DOI: 10.1038/nature04233

[2] Morozov SV, Novoselov KS, Katsnelson MI, Schedin F, Elias DC, Jaszczak JA, et al. Giant intrinsic carrier mobilities in graphene and its bilayer. Physical Review Letters. 2008;**100**(1):016602. DOI: 10.1103/PhysRevLett.100.016602

[3] Lee C, Wei XD, Kysar JW, Hone J. Measurement of the elastic properties of intrinsic strength of monolayer graphene. Science. 2008;**321**:385-388. DOI: 10.1126/science.1157996

[4] Zhang Y, Small JP, Amori MES, Kim P. Electric field modulation of galvanomagnetic properties of mesoscopic graphite. Physical Review Letters. 2005;**94**(17):176803. DOI: 10.1103/PhysRevLett.94.176803

[5] Kim KS, Zhao Y, Jang H, Lee SY, Kim JM, Kim KS, et al. Large-scale pattern growth of graphene films for stretchable transparent electrodes. Nature. 2009;**457**:706-716. DOI: 10.1038/nature07719

[6] Yang S-Y, Chang K-H, Lee Y-F, Ma C-CM HC-C. Constructing a hierarchical graphene-carbon nanotube architecture for enhancing exposure of graphene and electrochemical activity of Pt nanoclusters. Electrochemistry Communications. 2010;**12**(9):1206-1209. DOI: 10.1016/j.elecom.2010.06.020

[7] Ghislandi M, Tkalya E, Schilinger S, Koning CE, De With G. High performance graphene-and MWCNTs-based PS/PPO composites obtained via organic solvent dispersion. Composites Science and Technology. 2013;**80**:16-22. DOI: 10.1016/j.compscitech.2013.03.006

[8] Matos CF, Galembeck F, Zarbin F, Aldo JG. Multifunctional and environmentally friendly nanocomposites between natural rubber and graphene or graphene oxide. Carbon. 2014;**78**:469-479. DOI: 10.1016/j.carbon.2014.07.028

[9] Zhan Y, Lavorgna M, Buonocore G, Xia H. Enhancing electrical conductivity of rubber composites by constructing interconnected network of self-assembled graphene with latex mixing. Journal of Materials Chemistry. 2012;**22**(21):10464-10468. DOI: 10.1039/C2JM31293J

[10] Stankovich S, Dikin DA, Dommett GHB, Rohlhaas KM, Zimney EJ, Stach EA, et al. Graphene-based composite material. Nature. 2006;**442**:282-286. DOI: 10.1038/nature04969

[11] Nurhafizah MD. Magnetic properties of graphene oxide via a simple mixing with waste engine oil-based carbon nanotubes. SN Applied Sciences. 2020;**2**:534. DOI:10.1007/s42452-020-2361-8

[12] Shah RK, Hunter DL, Paul DR. Nanocomposites from poly (ethylene-co-methacrylic acid) ionomers: Effect of surfactant structure on morphology and properties. Polymer. 2005;**46**(8):2646-2662. DOI: 10.1016/j.polymer.2005.01.062

[13] Guo B, Tang Z, Zhang L. Transport performance in novel elastomer nanocomposites: Mechanism, design and control. Progress in Polymer Science. 2016;**61**:29-66. DOI: 10.1016/j.progpolymsci.2016.06.001

[14] Aguilar-Bolados H, Brasero J, Lopez-Manchado MA, Yasdani-

Pedram M. High performance natural rubber/thermally reduced graphite oxide nanocomposites by latex technology. Composites Part B: Engineering. 2014;**67**:449-454. DOI: 10.1016/j.compositesb.2014.08.010

[15] Zuberi M, Sherman DM, Cho Y. Carbon nanotube microspheres produced by surfactant-mediated aggregation. The Journal of Physical Chemistry C. 2011;**115**(10):3881-3887. DOI: 10.1021/jp110019e

[16] Mohamed A, Anas AK, Suriani AB, Ardyani T, Zin WMW, Ibrahim S, et al. Enhanced dispersion of multiwall carbon nanotubes in natural rubber latex nanocomposites by surfactants bearing phenyl groups. Journal of Colloid and Interface Science. 2015;**455**:179-187. DOI: 10.1016/j.jcis.2015.05.054

[17] Mohamed A, Ardyani T, Suriani AB, Brown P, Hollamby M, Sagisaka M, et al. Graphene-philic surfactants for nanocomposites in latex technology. Advances in Colloid and Interface Science. 2016;**230**:54-69. DOI: 10.1016/j.cis.2016.01.003

[18] Zhou S, Wei D, Shi H, Feng X, Xue K, Zhang F, et al. Sodium dodecyl benzene sulfonate functionalized graphene for confined electrochemical growth of metal/oxide nanocomposites for sensing application. Talanta. 2013;**107**:349-355. DOI: 10.1016/j.talanta.2013.01.041

[19] Yuan K, Xu Y, Uihlein J, Brunklaus G, Shi L, Heiderhoff R, et al. Straightforward generation of pillared, microporous graphene frameworks for use in supercapacitors. Advanced Materials. 2015;**27**(42):6714-6721. DOI: 10.1002/adma.201503390

[20] Jothi PR, Salunkhe RR, Pramanik M, Kannan S, Yamauchi Y. Surfactant-assisted synthesis of nanoporous nickel sulfide flakes and their hybridization with reduced graphene oxides for supercapacitor applications. RSC Advances. 2016;**6**(25):21246-21253. DOI: 10.1039/C5RA26946F

[21] Nurhafizah MD, Khor SY, Tan KL, Soga T. The synthesized reduced graphene oxide enhanced the capacitive behavior of activated carbon/pva as potential electrode materials. Journal of Nanostructures. 2019:**2**(10)1. DOI: 10.22052/jns.2019.196289.1919

[22] Nurhafizah MD, Suriani AB, Alfarisa S, Mohamed A, Isa IM, Kamari A, et al. The synthesis of graphene oxide via electrochemical exfoliation method. Advanced Materials Research. 2015;**1109**:55-59. DOI: 10.4028/www.scientific.net/amr.1109.55

[23] Nurhafizah MD, Suriani AB, Mohamed A, Soga T. Effect of voltage applied for graphene oxide/latex nanocomposite produced via electrochemical exfoliation and its application as conductive electrodes. Diamond and Related Materials. 2020;**101**:107624. DOI: 10.1016/j.diamond.2019.107624

[24] Nurhafizah MD, Aziz AA, Suriani AB, Mohamed A, Soga T. Low-temperature exfoliated graphene oxide incorporated with different types of natural rubber latex: Electrical and morphological properties and its capacitance performance. Ceramics International. 2020;**46**(5):5610-5622. DOI: 10.1016/j.ceramint.2019.11.005

[25] Smith RJ, Lotya M, Coleman JN. The importance of repulsive potential barriers for the dispersion of graphene using surfactants. New Journal of Physics. 2010;**12**:125008. DOI: 10.1088/1367-2630/12/125008

[26] Sun Z, Nicolosi V, Rickard D, Bergin SD, Aherne D, Coleman JN. Quantitative evaluation of surfactant-stabilized single-walled carbon nanotubes: Dispersion quality and its

correlation with zeta potential. Journal Physics Chemistry C. 2008;**112**:10629-10638. DOI: 10.1021/jp8021634

[27] Alemdar A, Gungor N, Erim FB. Effect of sodium dodecyl sulphate and sodium dodecyl benzene sulfonate on the flow behavior of purified bentonite dispersion. Journal of Materials Science Letters. 2003;**22**:89-90. DOI: 10.1023/A:1021836115359

[28] Cai X, Zhang Q, Wang S, Pen J, Zhang Y, Ma H, et al. Surfactant-assisted synthesis of reduced graphene oxide/polyaniline composites by gamma irradiation for supercapacitors. Journal of Materials Science. 2014;**49**(16):5667-5675. DOI: 10.1007/s10853-014-8286-0

[29] Mohamed A, Anas AK, Suriani AB, Azira AA, Sagisaka M, Brown P, et al. Preparation of multiwall carbon nanotubes (MWCNTs) stabilised by highly branched hydrocarbon surfactants and dispersed in natural rubber latex nanocomposites. Colloid & Polymer Science. 2014;**292**(11):3013-3023. DOI: 10.1007/s00396-014-3354-1

[30] Mohamed A, Trickett K, Chin SY, Cummings S, Sagisaka M, Hudson L, et al. Universal surfactant for water, oils, and CO2. Langmuir. 2010;**26**(17):13861-13866. DOI: 10.1021/la102303q

[31] Qiao SJ, Xu XN, Qiu Y, Xiao HC, Zhu YF. Simultaneous reduction and functionalization of graphene oxide by 4-hydrazinobenzenesulfonic acid for polymer nanocomposites. Nanomaterials. 2016;**6**(2):29. DOI: 10.3390/nano6020029

[32] Ferrari AC, Meyer JC, Scardaci V, Casiraghi C, Lazzeri M, Mauri F, et al. Raman spectrum of graphene and graphene layers. Physical Review Letters. 2006;**97**:187401. DOI: 10.1103/PhysRevLett.97.187401

[33] Graf D, Molitor FO, Ensslin K, Satmpfer C, Jungen A, Hierold C, et al. Spatially resolved Raman spectroscopy of single-and few-layer graphene. Nano Letters. 2007;**7**(2):238-242. DOI: 10.1021/nl061702a

[34] Nurhafizah Binti Md Disa. Synthesis of graphene oxide using electrochemical exfoliation method for electrode materials application. thesis dissertation; 2017

[35] Paredes JI, Burghard M, Martanez-Alonso TJMD. Graphitization of carbon nanofibers: Visualizing the structural evolution on the nanometer and atomic scales by scanning tunneling microscopy. Applied Physics A. 2005;**80**(4):675-682. DOI: 10.1007/s00339-004-3109-9

[36] Kazi SN, Badarudin A, Zubir MNM, Ming HN, Misran M, Sadeghinezhad E, et al. Investigation on the use of graphene oxide as novel surfactant to stabilize weakly charged graphene nanoplatelets. Nanoscale Research Letters. 2015;**10**(1):1-15. DOI: 10.1186/s11671-015-0882-7

[37] Suriani AB, Nurhafizah MD, Mohamed A, Zainol I, Masrom AK. A facile one-step method for graphene oxide/natural rubber latex nanocomposite production for supercapacitor applications. Materials Letters. 2015;**161**:665-668. DOI: 10.1016/j.matlet.2015.09.050

[38] Suriani AB, Nurhafizah MD, Mohamed A, Masrom AK, Mamat MH, Malek MF, et al. Electrical enhancement of radiation-vulcanized natural rubber latex added with reduced graphene oxide additives for supercapacitor electrodes. Journal of Materials Science. 2017;**52**:6611-6622. DOI: 10.1007/s10853-017-0897-9

[39] Potts JR, Shankar O, Du L, Ruoff RS. Processing-morphology-property relationships and composite theory analysis of reduced graphene oxide/natural rubber nanocomposites.

Macromolecules. 2012;**45**(15):6045-6055. DOI: 10.1021/ma300706k

[40] Clark MD, Subramanian S, Krishnamoorti R. Understanding surfactant aided aqueous dispersion of multi-walled carbon nanotubes. Journal of Colloid and Interface Science. 2011;**354**(1):144-151. DOI: 10.1016/j.cis.2010.10.027

[41] Ghasemi S, Hosseinzadeh R, Jafari M. MnO_2 nanoparticles decorated on electrophoretically deposited graphene nanosheets for high performance supercapacitor. International Journal of Hydrogen Energy. 2015;**40**(2):1037-1046. DOI: 10.1016/j.ijhydene.2014.11.072

[42] Graydon JW, Panjehshashi M, Kirk DW. Charge redistribution and ionic mobility in the micropores of supercapacitors. Journal of Power Sources. 2014;**245**:822-829. DOI: 10.1016/j.jpowsour.2013.07.036

[43] Andreas HA. Self-discharge in electrochemical capacitors: A perspective article. Journal of The Electrochemical Society. 2015;**162**(5):A5047. DOI: 10.1149/2.0081505jes

[44] He Y, Zhang Y, Li X, Lv Z, Wang X, Liu Z, et al. Capacitive mechanism of oxygen functional groups on carbon surface in supercapacitors. Electrochimica Acta. 2018;**282**:618-625. DOI: 10.1016/j.electacta.2018.06.103

[45] Kshetri T, Tran DT, Nguyen DC, Kim NH, Lau K, Lee JH. Ternary graphene-carbonnanofibers-carbonn anotubesstructureforhybridsupercap acitor. Chemical Engineering Journal. 2020;**380**:122543. DOI: 10.1016/j.cej.2019.122543

[46] Li T, Wang X, Liu P, Yang B, Diao S, Gao Y. Synthesis of graphene/polyaniline copolymer for solid-state supercapacitor. Journal of Electroanalytical Chemistry. 2020;**860**:113908. DOI: 10.1016/j.jelechem.2020.113908

[47] Layek RK, Uddin ME, Kim NH, Tak Lau AK, Lee JH. Noncovalent functionalization of reduced graphene oxide with pluronic F127 and its nanocomposites with gum Arabic. Composites Part B: Engineering. 2017;**128**:155-163. DOI: 10.1016/j.compositesb.2017.07.010

Chapter 5

An Electrochemical Sensor Based on Electroreduction of Graphene Oxide on a Glassy Carbon Electrode Using Multiple Pulse Amperometry for Simultaneous Determination of L-Dopa and Benserazide

Thiago Gabry Barbosa, Ana Elisa Ferreira Oliveira and Arnaldo César Pereira

Abstract

In this work, we described the development an electrochemical sensor based on electroreduction of graphene oxide (rGO) on a glassy carbon electrode (GCE) for simultaneous determination of L-Dopa and benserazide. For the elaboration of the GCE/rGO, the developed methodology was based on the electrochemical technique: multiple pulse amperometry (MPA). The MPA was more stable and efficient for the formation of rGO film, under optimum conditions (pH 6.00; concentration of rGO 2.00 mg mL^{-1}; time 450 s; potentials −0.60, −0.70, −0.80, −0.90, −0.95, −1.00, −1.10, −1.20, and − 1.30 V). After the film was formed, the cyclic voltammetry was used to detect LD and BZ in real samples and optimized conditions 0.05 mol L^{-1} PBS (pH 5.50). The linear range for the LD is 25–425 μmol L^{-1} and the BZ of 5–80 μmol L^{-1}. The limit of detection calculated was 17.10 (LD) and 2.99 (BZ) μmol L^{-1}.

Keywords: graphene oxide, electroreduction, simultaneous determination, L-Dopa, benserazide

1. Introduction

Parkinson's disease (PA) is an illness that affects about 1% of the world population, according to the World Health Organization (WHO), and this number tends to increase considerably, as demonstrated by recent studies conducted by the University of Rochester. Its estimated that the number of people affected by the PA in the 15 countries analyzed will more than double up to 2040 [1]. Thus, it becomes extremely important to develop research on this disease and the treatment employed. Mainly alternative routes for the quantification of compounds are present in the drug used in the treatment of BP, which provides an effective and lower cost treatment.

Parkinson's disease is a condition that attacks the central nervous system (CNS) and the brain, and this evil affects the amount of dopamine in the body. Dopamine (DA) is an existing neurotransmitter catecholamine in the CNS and in the mammalian brain. It has a vital role in maintaining the activities of the central nervous system, cardiovascular system, and hormonal system [2]. Patients suffering from BP have a considerable decay in the production OF DA in the brain. The medications for this disease have the design of elevating the dopamine index in the brain [3].

However, it is not possible to inject dopamine directly into the patient, because the blood-brain barrier does not allow the arrival of this hormone in the encephalon. With this, L-Dopa (LD) is used, a medicine that can overcome such a barrier and is converted into the encephalon in DA [4, 5]. Despite this, when there is an irregularity in the levels OF DA, there are some side effects such as nausea, vomiting, cardiac arrhythmia, schizophrenia, and dyskinesia [2, 6].

Unfortunately, Dopa decarboxylase (DDC) quickly converts LD into the bloodstream, so only a small percentage reaches the brain. By inhibiting the enzyme, higher amounts of L-Dopa administered can reach the brain [7]. In order not to be converted "early," the drug hydrochloride is introduced in the benserazide (BZ), which acts as a DDC inhibitor [8].

Electroanalytical methods emerged as an alternative line when compared to chromatographic methods that demand the use of expensive instruments, which are experimentally complex and require a lot of time to analyze with various pretreatments of the sample. These methods are based on the different electroanalytical and electrochemical techniques. These in turn have excellent sensitivity, selectivity, speed, reduced costs, and possibility of miniaturization of the system, making the electrochemical sensors a promising tool to supplement the existing techniques. Therefore, the development of new strategies with the aim of perfecting and improving the electrochemical techniques is promising [9–11].

Solid electrodes of different materials are employed in the electroanalytical methods; one of the most widely used materials is those based on carbonaceous materials, such as carbon graphite, pyrolytic carbon, and glass carbon. In addition to serving as a base electrode, carbonaceous materials are also used as modifiers, in order to catalyze the redox process and/or increase selectivity. Some examples that can be found in the literature are the use of reduced graphene oxide (RGO), graphene, nanotubes (single or multiple wall), etc.

For this work the glass carbon electrode (GCE) was chosen as the base electrode and the reduced graphene oxide (rGO) by electroreduction. The GO could be reduced also by thermal and chemical process; however the electrochemical reduction causes a homogeneity and stability of the surface of the work electrode; for this reason we opted for the use of electroreduction of GO.

The process of reduction of GO by electrochemical route is described in the literature in two ways, one using conventional amperometry (Amp) and the other using cyclic voltammetry (CV). The methodology employing the Amp consists in the application of a single potential in a specific time range, already the methodology employing CV focused on the application of a potential window at different scanning speeds and number of cycles.

In order to add value to this work, the use of multiple pulse amperometry (MPA) was tested, a technique derived from conventional amperometry, and there are still no reports in the literature of this technique used for the purpose of reducing GO. Since MPA consists in the application of different potentials in a certain time range, this process has as an advantage the application of activation potentials of the electrode surface and cleaning [12]. Therefore, these possibilities can generate better efficiency in the reduction process; the number of pulses of potential applicable considering the software GPES reaches 10 and has the possibility of acquiring the current as a function of time to each potential pulse.

The glass carbon electrode has distinct properties of its allotropic forms, pyrolytic graphite, and diamond carbon, which are also used in electrochemistry. However, it presents the best synergism of physical properties, affinity with the modifier material, and its good affinity with the selected immobilization methodology, which is a prerequisite described in the literature [13].

The material that will carry out the chemical modification of the base electrode will be the GO, which will be reduced electrochemically in order to make it a conductive material, thus decreasing the resistivity of the electrode facilitating the transfer of electrons. The GO consists of a graphene sheet, a carbon structure consisting of sp^2 bonds, which has these connections transformed into sp^3 bonds by the substitutes groups, which removes its conductive characteristic. The reduction process causes this material to lose some functional groups and return to conductive characteristics and maintain the interaction from the functional groups [14–17].

The proposal of this work is the development of an electroanalytical method for the electrochemical reduction of graphene oxide and the simultaneous determination of LD and BZ in an electrochemical cell of three electrodes.

2. Construction of an electrochemical sensor based on electroreduction of graphene oxide on a GCE

2.1 Instrumentation

Electrochemical measurements and the formation of the work electrode modifier film were performed in a multi-potentiostat/galvanostat model PGSTAT101 coupled to a microcomputer containing the new 1.11 software and a microcomputer containing the GPES 4.1 software. As a system for obtaining electrochemical measurements, an electrochemical cell containing three electrodes was used: a saturated Ag/AgCl electrode containing 3.0 mol L^{-1} of KCl as a reference electrode, a platinum wire as an auxiliary electrode, and the GCE/rGO as a working electrode.

Sodium phosphate dibasic heptahydrate, monobasic sodium phosphate monohydrate obtained from the Synth®, phosphoric acid, and 3.4-dihydroxy-L-phenylalanine were used; benserazide hydrochloride is acquired from Sigma-Aldrich®. All the chemical reagents used were analytical grade, and the buffer solutions were prepared with purified water by Millipore Milli-Q system.

The solution containing L-Dopa was prepared with a solution of phosphoric acid pH 2.00 and concentration 0.1 mol L^{-1}, and the solution of benserazide was prepared in phosphate buffer pH 5.50 and concentration 0.1 mol L^{-1}.

2.2 Pretreatment of glassy carbon electrode

The glassy carbon electrode was pretreated by mechanical polishing with Alumina, HNO_3, and H_2O and also by electrochemical activation performed in a 0.5 mol L^{-1} H_2SO_4 solution for 20 scans at scan rate of 100 MV s^{-1} and a potential range −0.2 to 1.6 V.

2.3 Electroreduction of graphene oxide

The graphene oxide (GO) was synthesized based on previous works published [14]. The dispersion of GO was prepared in ethanol and Nafion 5% v/v. The construction of the proposed sensor was carried out by adding 5 μL of the GO suspension

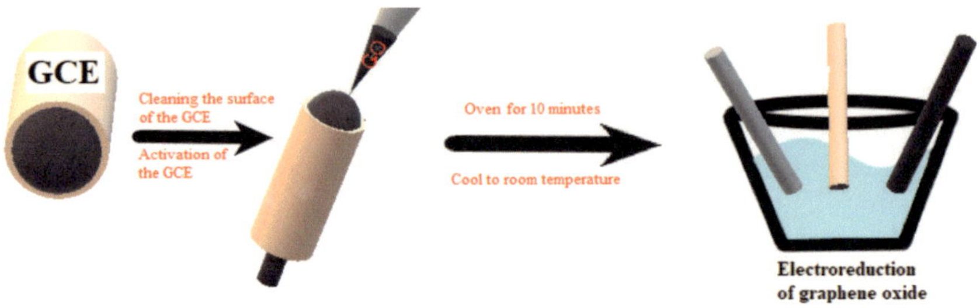

Figure 1.
Process of modifying the surface of the glassy carbon electrode.

in the GCE. Then the electrode was kept in the oven at 60°C for 10 minutes. The electrochemical reduction of the GO was performed at multiple pulses to obtain the rGO. The reduction was carried out using a three-electrode setup: reference electrode (Ag/AgCl), auxiliary electrode (platinum wire), and working electrode (GCE) at PBS 0.1 mol L^{-1} and pH 7.00. **Figure 1** summarizes the process of modifying of the GCE.

2.4 Multiple pulse amperometry

The electrochemical reduction of rGO by MPA is not described in the literature. However, based on the procedure via amperometry and by cyclic voltammetry [18–22], some conditions were adopted as a starting point. According to the literature review, [22] the application of a negative potential after the reduction process using cyclical voltammetry tends to improve the reduction and homogeneity of the film, with this in mind and in order to enjoy the best performance possible of the MPA technique was adopted 9 potentials for the process of reduction of GO, optimizing which would be these potentials, the time of application and the concentration of GO.

Considering that mechanisms may change according to the conditions used in electroreduction, some experimental parameters will be evaluated in the MPA reduction process. Initially, four parameters were optimized—applied potential, pulse application time, GO concentration, and pH effect. The objective of the optimization of these parameters is to develop a sensor with higher sensitivity in the determination of LD and BZ, and the conditions and results obtained in the optimization process will be discussed later.

3. Simultaneous determination of L-Dopa and benserazide

3.1 Electrochemical behavior of L-Dopa and benserazide

The electrochemical behavior of LD against different sensor configurations, bare GCE and GCE/rGO before being optimized and after optimization, is shown in **Figure 2A**. An associated oxidation peak around 0.45 V is observed from the oxidation of hydroxyls bound to the LD aromatic ring. This oxidation process involves two electrons, and it is a similarly reversible process [23–25]. The cyclic voltammogram obtained with GCE/rGO after optimizing exhibits a current variation of approximately 1300% over simple GCE, as shown in **Figure 2B**. The observed current gain for L-Dopa is also observed in benserazide, thus making rGO a viable material for GCE modification.

The increase in current observed in the oxidation process of LD and BZ is due to some characteristics of rGO. Reduced graphene oxide consists of a graphene sheet with the presence of some functional groups along its structure, which will make the interaction between the base electrode and the analyte easier. In addition, rGO is a conductive material thus facilitating the exchange of electrons between the analyte and the WE.

It is noteworthy that the increase in observed current from the addition of rGO leads to an increase in the linear working range of the sensor and an increase in its sensitivity.

3.2 Optimization of experimental parameters

3.2.1 Effect of applied potential range

Initially we analyzed potential range to reduce the graphene oxide. These potentials were in the window of −0.50 to −1.40 V. **Figure 3** indicates the behavior of the analytes before the different potentials adopted, and the anodic current was adopted as the selection method of the best result and also took into account the separation of current peaks anodic of LD and BZ. This figure does not represent the process of reducing but the electrochemical behavior of analytes using the GCE/rGO after the reduction.

Among the potentials adopted, (−0.50, −0.60, −0.70, −0.80, −0.90, −1.00, −1.10, − 1.20, −1.30), (−0.60, −0.70, −0.80, −0.90, −0.95, − 1.00, −1.10, − 1.20, −1.30), and (−0.70, −0.80, −0.90, −1.00, −1.05, −1.10, −1.20, −1.30, −1.40), the

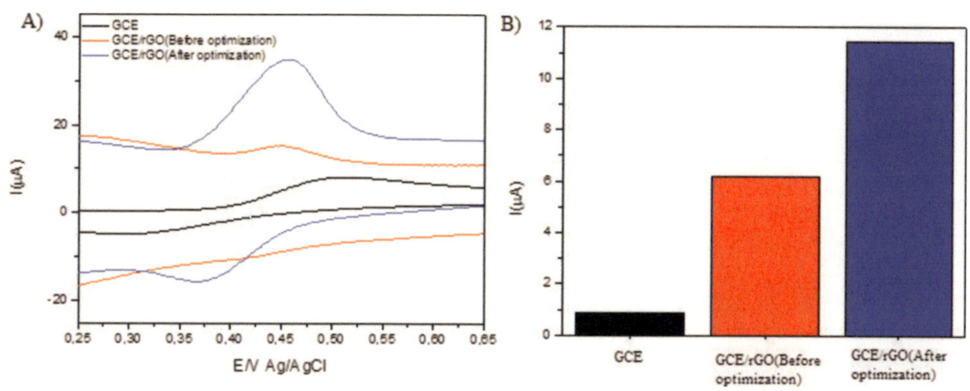

Figure 2.
LD 0.1 µmol L^{-1} electrochemical behavior in the analytical response (A) GCE, GCE/rGO before optimization, and GCE/rGO after optimization (B).

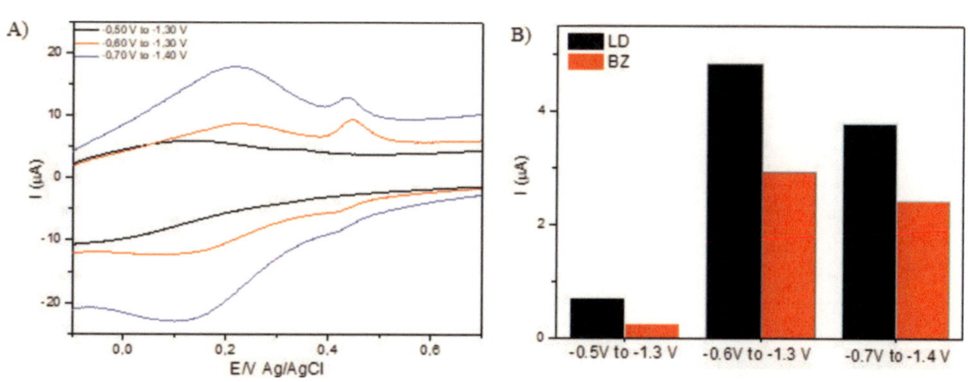

Figure 3.
(A) Cyclic voltammograms obtained at different electrodes prepared via MPA with different potentials (PBS 0.1 mol L^{-1}, pH 5.50) after addition of 1.0 × 10^{-5} mol (BZ) and 9.95 × 10^{-5} mol L^{-1} (LD). (B) Analytical response.

range of −0.60 to −1.30 V presented the best result; therefore this value was adopted as the optimum value, and the analysis process continued.

3.2.2 Effect of time

After determining the potentials to be applied in the electroreduction, the time of application of these potentials was evaluated. Three application times were analyzed: 400, 450, and 500. **Figure 4** represents the voltammograms obtained after the GO reduction process, and this analysis was performed in the presence and absence of analyte for electrodes prepared at different times of reduction.

Through the current difference, it was determined which would be the best fear of the application of the pulses, having as optimum time 450 s.

3.2.3 Effect of GO concentration

With the potentials and time of application well-defined, the evaluation of the GO concentration added to the electrode surface was performed, from 0.5 to 3 mg ml^{-1}. **Figure 5** shows the voltammograms for the different concentrations of the solutions added on the GCE, and the optimum condition was defined by means of the current difference. This condition was 2.00 mg mL^{-1}.

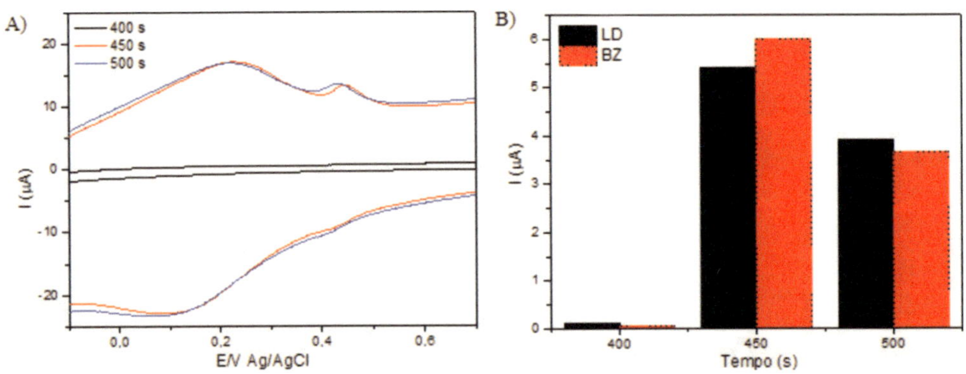

Figure 4.
(A) Cyclic voltammograms obtained at different electrodes prepared via MPA with different times (PBS 0.1 mol L^{-1}, pH 5.50) after addition of 1.0×10^{-5} mol (BZ) and 9.95×10^{-5} mol L^{-1} (LD). (B) Analytical response.

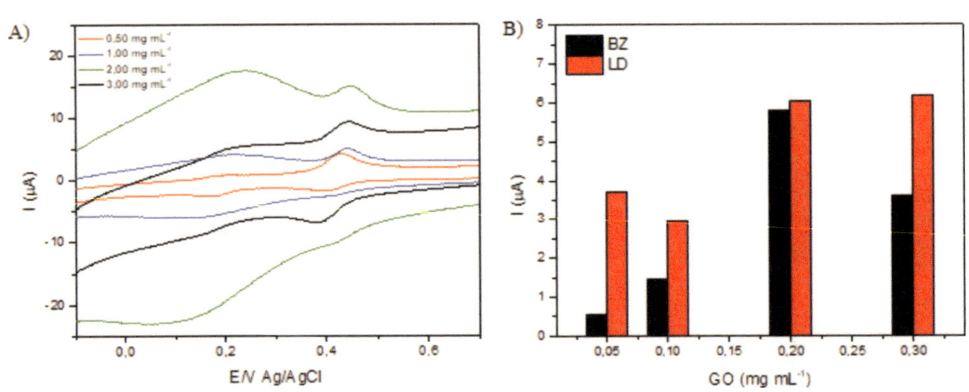

Figure 5.
(A) Cyclic voltammograms obtained at different electrodes prepared via MPA with different GO concentrations (PBS 0.1 mol L^{-1}, pH 5.50) after addition of 1.0×10^{-5} mol (BZ) and 9.95×10^{-5} mol L^{-1} (LD). (B) Analytical response.

3.2.4 Effect of pH

Finally, the pH influence of the electrolytic solution in the reduction process was evaluated, since it can cause interferences in the process of electroreduction, such as damage to the modifier material. Concerning this study, the optimal measurements previously evaluated in **Figure 6** can be observed, and the influence of the pH value of the support solution—phosphate buffer at 0.10 mol L^{-1}—in the formation of the rGO film. By means of the current difference, the best pH value was defined, which was 6.00.

It should be noted that when the dispersion concentration of the modified material was optimized due to the influence of the thickness of the modification and the number of functional groups, and the thicker and the smaller the number of groups, the lower the number of functional loads will be transported. The potential is applied to eliminate the excess of functional groups, transforming the GO into or returning to give conductive characteristics of this material. The reduction time in turn allows a better interaction of the film with the reduction process making its surface more homogeneous.

Therefore, it is extremely important that there is an optimization of the parameters of the technique and the concentration of modifier material to obtain a more efficient sensor, that is, with good sensitivity and selectivity.

3.3 Optimization of experimental conditions

After the optimization of the GO reduction process, it was evaluated which of these presented better conditions for the continuation of the analyses of LD and BZ. The choice of the best technique was based on the separation of the peaks of LD and BZ and the higher anodic current.

3.3.1 Influence of ionic force

It is known that the ionic force is directly linked to the feasibility of the analysis once the concentration of the ions influences in the transport of loads, which can generate an increase in the signal of the analyte.

The determination of the ionic strength of the solution was made by means of the difference of white current and the addition of the analytes obtained in a buffer system pH 5.50, and the electrolyte concentration support was varied from 0.025 to 0.200 mol L^{-1}. **Figure 7** illustrates the voltammograms obtained for these different concentrations, and through them it was possible to define that the ionic force that

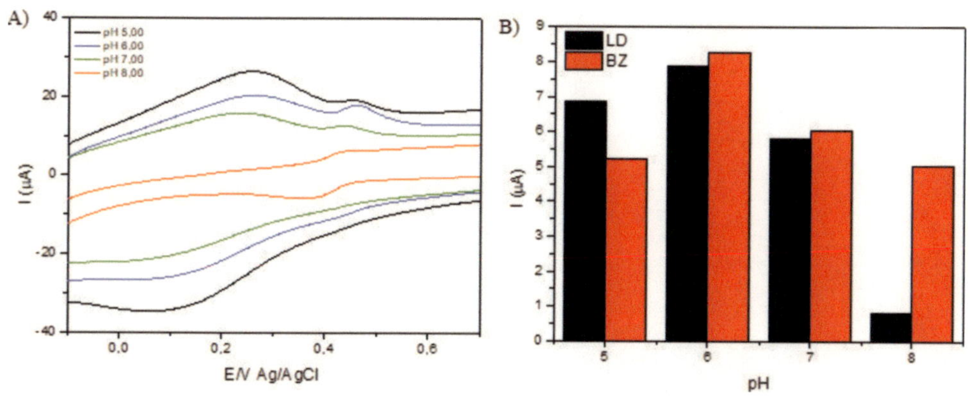

Figure 6.
(A) Cyclic voltammograms obtained at different electrodes prepared via MPA with different pH (PBS 0.1 mol L^{-1}, pH 5.50) after addition of 1.0×10^{-5} mol (BZ) and 9.95×10^{-5} mol L^{-1} (LD). (B) Analytical response.

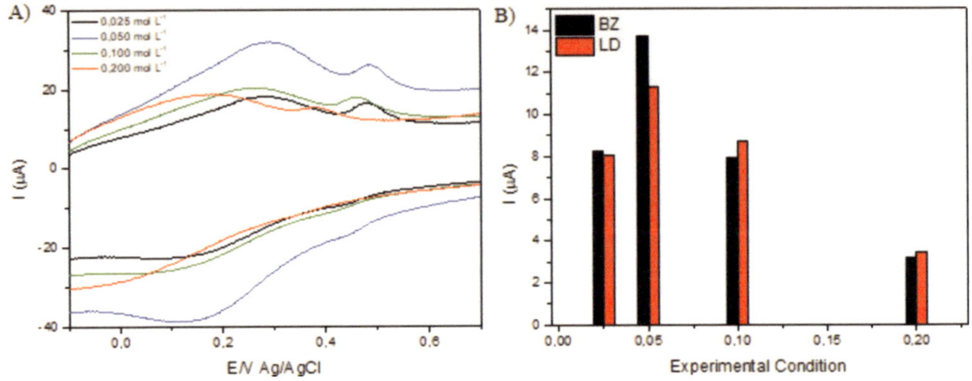

Figure 7.
(A) Cyclic voltammograms obtained at different electrodes prepared via MPA with different electrolyte concentration (PBS pH 5.50) after addition of 1.0×10^{-5} mol (BZ) and 9.95×10^{-5} mol L^{-1} (LD). (B) Analytical response.

best suited was 0.050 mol L^{-1}. The higher concentration may be happening with the competition between the electrolyte ions support and the analytes and concentrations below these there are not enough ions to carry out the transport of loads.

3.3.2 pH influence

Subsequently, the pH effect was evaluated, which may influence the electrode stability and provide secondary reactions of the analytes, such as LD, which in basic ph suffers secondary reactions which degrade it. This parameter was also evaluated by analyzing anodic current, and the higher these differences are the signal obtained, **Figure 8** illustrates the voltammograms for the different pH values of the electrolyte support, the optimum value for analysis was determined being 5.50.

3.4 Calibration curve

First, we calculated the amount of analyte needed to prepare a solution containing LD and BZ in the same proportion found in Prolopa®, from 5.23 mol of LD to 1.00 mol of BZ, which would be analyzed. Through this solution, the voltammetry analysis was performed using proposed sensor (previously optimized) as shown in **Figure 9**. The analytical curves were constructed for LD (**Figure 10**) and BZ (**Figure 11**).

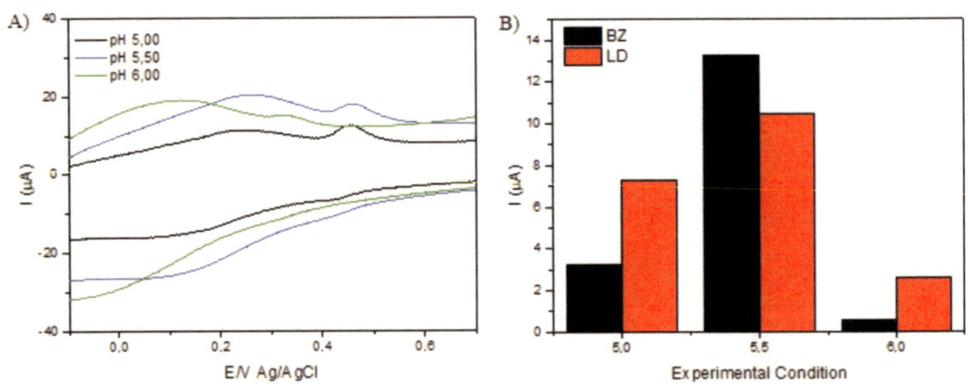

Figure 8.
(A) Cyclic voltammograms obtained at different electrodes prepared via MPA with different pH of the electrolyte (PBS 0.050 mol L^{-1}) after addition of 1.0×10^{-5} mol (BZ) and 9.95×10^{-5} mol L^{-1} (LD). (B) Analytical response.

Having all the data optimized, it was possible to construct the final analytical curves by means of successive additions of LD 5.08 mmol L^{-1} and BZ 0.97 mmol L^{-1} in order to obtain the sensibility, the detection limit (LOD), and the limit of quantification (LOQ) for each analyte.

The sensibility for LD was 0.05148 μmol L^{-1} and 0.10303 μmol L^{-1} for BZ. For the analytical curve, to have an acceptable degree of reliability, the value of R must be close to 1. The values obtained for R are above 0.98, which indicates a considerable good degree of reliability.

The detection limit (LOD) represents the smallest amount of the analyte present in a sample that can be detected by the method. From the parameters of the analytical curve, the LOD can be expressed by LOD = 3 × (S/B), where S is the standard deviation of the white and B is the slope of the analytical curve.

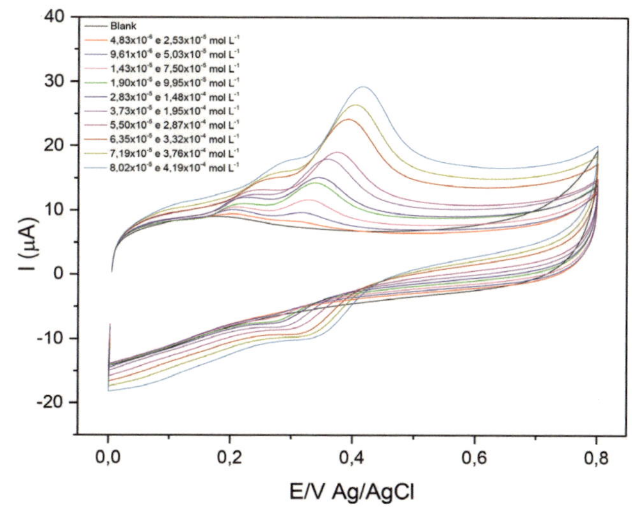

Figure 9.
Voltammograms (PBS 0.05 mol L^{-1} pH 5.5) and calibration curve, using cyclic voltammetry, containing LD in the concentration range of 2.53 × 10^{-5} mol L^{-1} to 4.19 × 10^{-4} mol L^{-1} and BZ of 4.83 × 10^{-6} mol L^{-1} to 8.02 × 10^{-5} mol L^{-1}.

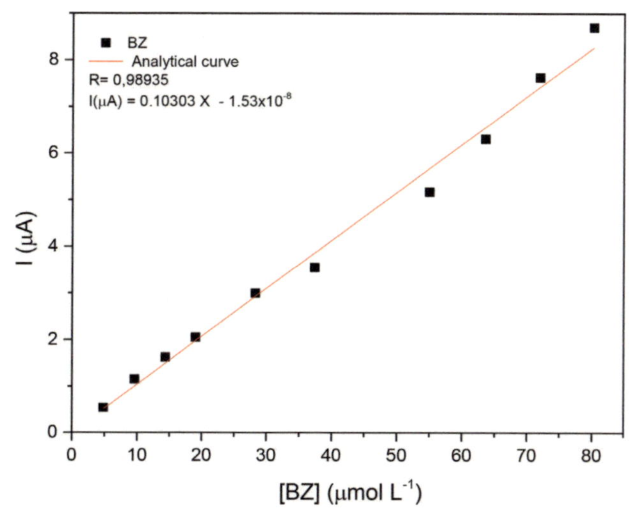

Figure 10.
Calibration curve of BZ in phosphate buffer, pH 5.50, and concentration 0.05 mol L^{-1}.

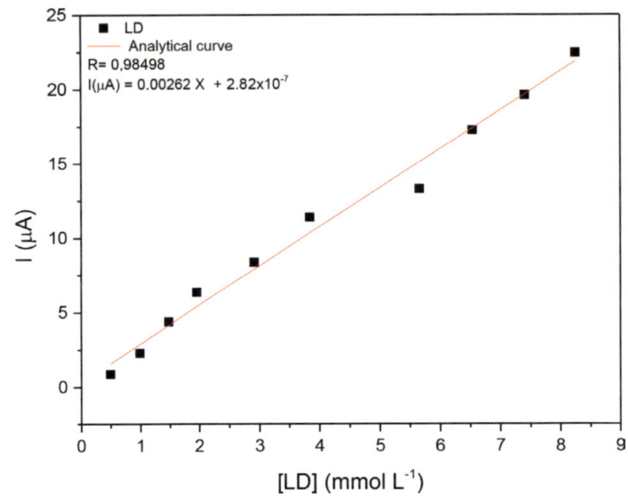

Figure 11.
Calibration curve of LD in phosphate buffer, pH 5.50, and concentration 0.05 mol L^{-1}.

The limit of quantification (LOQ) is the smallest amount of the analyte in a sample that can be determined with accuracy and accuracy acceptable under the established experimental conditions. The LOQ is established through the analysis of samples containing decreasing concentrations of the analyte up to the lowest determinable level and can be expressed by the equation: LOD = 10 × (S/B), where S and B have the same values found for the LOD, previously.

The limit of detection and the limit of quantification of LD and BZ obtaining was, respectively, 5.14 μmol L^{-1} and 17.10 μmol L^{-1} for LD and 8.96×10^{-1} μmol L^{-1} and 2.99 μmol L^{-1} and BZ. However, the detection limit can be adopted as the first point of the analytical curve, thus becoming 4.83 μmol L^{-1} for BZ and 25.30 μmol L^{-1} for LD.

After the calibration curve was constructed, reproducibility and repeatability tests were performed. Reproducibility was confirmed by inter-day testing, where analyses of the same sample were performed on different days, and by intra-day testing, where five electrodes were built and samples of the same concentration were analyzed, after these tests. A good reproducibility must have standard deviation of less than 5% which will occur.

Repeatability can be ratified by inter- and intra-day analysis, and it was made by intra-day analysis, where an electrode was built to analyze five samples of the same concentration, which presented a standard deviation below 5%, which statistically proves the repeatability of the method.

It is noteworthy that all analyses presented throughout the text were made in triplicate, where a relative error of less than 5% was observed, which statistically proves that the analyses presented are within an acceptable standard.

3.5 Application in pharmaceutical formulation

In order to evaluate the applicability of the proposed method, a solution of LD and BZ was prepared employing the sample of the drug (Prolopa®) at the concentration of 5.08 mmol L^{-1} of LD and 0.97 mmol L^{-1} of BZ, having the nominal value of each analyte present in the Prolopa® as a reference for calculating the indicated concentrations. This solution was analyzed by the proposed sensor.

Then, the recovery tests were performed. The results were obtained in triplicate and are presented in **Tables 1** and **2**.

Measure	[LD] added	[LD] recovered	Recovery (%)	E_R (%)
1	2.53×10^{-5} mol L^{-1}	2.60×10^{-5} mol L^{-1}	102.77	2.77
2	2.87×10^{-4} mol L^{-1}	2.78×10^{-4} mol L^{-1}	96.86	−3.14
3	4.19×10^{-4} mol L^{-1}	3.99×10^{-4} mol L^{-1}	97.91	−2.09

Table 1.
Results of the LD addition and recovery tests in a solution prepared with the drug.

Measure	[BZ] added	[BZ] recovered	Recovery (%)	E_R (%)
1	4.83×10^{-5} mol L^{-1}	4.95×10^{-5} mol L^{-1}	102.48	2.48
2	5.50×10^{-5} mol L^{-1}	5.25×10^{-5} mol L^{-1}	95.45	−4.55
3	8.02×10^{-5} mol L^{-1}	8.41×10^{-5} mol L^{-1}	104.86	4.86

Table 2.
Results of the addition and recovery tests of BZ in a solution prepared with the drug.

Through the results presented in **Tables 1** and **2**, it was possible to observe that the sensor proved to be promising in the simultaneous determination of the analytes with a maximum relative error of ±5%. And the results found were analyzed in the same system with successive additions of the drug solution.

It is noteworthy that the methodology of modification of the surface of the GCE is rapid and that the data obtained when compared with those present in the literature [23, 26–33] has performance as good as the sensors described; however, the preparation methodology of some is more laborious than the proposal in this work.

4. Conclusion

The proposed methodology was successfully developed, since it proved to be effective in the simultaneous determination of LD and BZ. It is worth noting that the methodology employed uses a rapid analysis technique, when comparing the separation techniques (chromatography).

In short, the work proved to be efficient and innovative when compared to the literature but can have the methodology of reduction by MPA improved. Thus, this study enable an improvement in the methodology of analysis and the description of this technique with the purpose of electroreduction and/or electropolymerize in the literature.

Acknowledgements

We thank FAPEMIG, CAPES, CNPQ and Rede Mineira de Química for the continuous support of our research.

Author details

Thiago Gabry Barbosa, Ana Elisa Ferreira Oliveira and Arnaldo César Pereira*
Departamento de Ciências Naturais, Universidade Federal de São João del-Rei, UFSJ, São João del-Rei, MG, Brazil

*Address all correspondence to: arnaldocsp@yahoo.com.br

References

[1] Dorsey ER, Sherer T, Okun MS, Bloem RB. The emerging evidence of the Parkinson pandemic. Journal of Parkison's Disease. 2018;**8**(Suppl 1):S3-S8. DOI: 10.3233/JPD-181474

[2] Venton BJ, Wightman RM. Psychoanalytical electrochemistry: Dopamine and behavior. Analytical Chemistry. 2003;**75**:414A-421A. DOI: 10.1021/ac031421c

[3] Olanow CW, Obeso JA, Stocchi F. Continuous dopaminereceptor treatment of Parkinson's disease: Scientific rationale and clinical implications. Lancet Neurology. 2006;**5**:677-687. DOI: 10.1016/S1474-4422 (06) 70521-X

[4] Hershey T, Black KJ, Carl JL, Mcgee-Minnich L, Snyder AZ. Long term treatment and disease severity change brain responses to levodopa in Parkinson's disease. Journal of Neurology, Neurosurgery, and Psychiatry. 2003;**74**:844-851. DOI: 10.1136/jnnp.74.7.844

[5] Khor SP, Hsu A. The pharmacokinetics and pharmacodynamics of levodopa in the treatment of Parkinsons disease. Current Clinical Pharmacology. 2007;**2**(3):234-243. DOI: 10.2174/157488407781668802

[6] Rezaei B, Shams-Ghahfarokhi L, Havakeshian E, Ensafi AA. An electrochemical biosensor based on nanoporous stainless steel modified by gold and palladium nanoparticles for simultaneous determination of levodopa and uric acid. Talanta. 2016;**158**:42-50. DOI: 10.1016/j.talanta.2016.04.061

[7] Burkhard P, Dominici P, Borri-Voltattorni C, Jansonius JN, Malashkevich VN. Structural insight into Parkinson's disease treatment from drug-inhibited DOPA decarboxylase. Nature Structural Biology. 2001;**8**(11):963-967. DOI: 10.1038/nsb1101-963

[8] Treseder SA, Rose S, Summo L, Jenner P. Commonly used Lamino acid decarboxylase inhibitors block monoamine oxidase activity in the rat. Journal of Neural Transmission. 2003;**110**(3):229-238. DOI: 10.1007/s00702-002-0778-4

[9] Norouzi P, Larijani B, Faridbod F, Ganjali MR. A novel method for ultra trace measurement of bentazon based on nanocomposite electrode and continuous coulometric FFTCyclic voltammetry. International Journal of Environmental Research. 2015;**9**:101-108

[10] Wang Y, Hasebe Y. Uricase-adsorbed carbon-felt reactor coupled with a peroxidase-modified carbon-felt-based H_2O_2 detector for highly sensitive amperometric flow determination of uric acid. Journal of Pharmaceutical and Biomedical Analysis. 2012;**57**:125-132. DOI: 10.1016/j.jpba.2011.08.021

[11] Trojanowicz M. Recent developments in electrochemical flow detections—A review: Part II. Liquid chromatography. Analytica Chemica Acta. 2011;**688**:8-35. DOI: 10.1016/j.aca.2010.12.024

[12] de Miranda JAT. Simultaneous determination of aspirin and ascorbic acid in medications using flow injection analysis with amperometric detection [master's thesis]. Uberlândia: Federal University of Uberlândia—Graduate Program in Chemistry; 2011. DOI: 10.1002/elan.200804262

[13] Pereira AC. Electrochemical behavior of some electroactive organic compounds immobilized in inorganic media aiming at the development of sensors for NADH [PhD thesis]. Unicamp: Institute of Chemistry; 2003

[14] Oliveira AEF, Braga GB, Tarley CRT, Pereira AC. Thermally reduced graphene oxide: Synthesis, studies and characterization. Journal of Materials Science. 2018;**53**:12005-12015. DOI: 10.1007/s10853-018-2473-3

[15] Li Y, Martens I, Cheung KC, Bizzotto D. Electrodeposition of reduced graphene oxide onto gold electrodes: Creating thin electrochemically active and optically transparent overlayers. Electrochimica Acta. 2019;**319**:649-656. DOI: 10.1016/j.electacta.2019.07.004

[16] Yao SY, Cai WL, Liu L, Liao XQ, Tao KL, Feng F, et al. Electrochemical behavior of eriocitrin and highly sensitive determination based on an electrochemically reduced graphene oxide modified glassy carbon electrode. Analytical Methods. 2016;**18**:3722-3729. DOI: 10.1039/C6AY00064A

[17] Wang X, Shen W, Zhang X, Guo S, Gao Y, Li X, et al. Indirect electrochemical determination of ribavirin using boronic acid-diol recognition on a 3-aminophenylboronic acid-electrochemically reduced graphene oxide modified glassy carbon electrode (APBA/ERGO/GCE). Analytical Letters. 2019;**52**:1-14. DOI: 10.1080/00032719.2019.1576716

[18] Motta R, Schrebler R, Zanoni R, Dalchiele EA. Insights from experiment and theory into the electrochemical reduction mechanism of graphene oxide. Electrochimica Acta. 2019;**304**:231-238. DOI: 10.1016/j.electacta.2019.02.108

[19] de Camargo MNL et al. Tuning the electrochemical reduction of graphene oxide: Structural correlations towards the electrooxidation of nicotinamide adenine dinucleotide hydride. Electrochimica Acta. 2016;**197**:194-199. DOI: 10.1016/j.electacta.2015.09.022

[20] Kumar DR, Kesavan S, Bayonosa ML, Shin JJ. 3,5-Diamino-1,2,4-triazole@electrochemically reduced graphene oxide film modified electrode for the electrochemical determination of 4-nitrophenol. Electrochimica Acta. 2017;**246**:1131-1140. DOI: 10.1016/j.electacta.2017.06.116

[21] Palakollu VN et al. Electrochemically reduced graphene oxide/poly-glycine composite modified electrode for sensitive determination of L-dopa. Materials Science and Engineering: C. 2017;**77**:394-404. DOI: 10.1016/j.msec.2017.03.173

[22] García-Argumáneza A, Llorentea I, Caballero-Calerob O, Gonzálezc Z, Menéndezc R, Escuderoa ML, et al. Electrochemical reduction of graphene oxide on biomedical grade CoCr alloy. Applied Surface Science. 2019;**465**:1028-1036. DOI: 10.1016/j.apsusc.2018.09.188

[23] Sunder GSS, Rohanifar A, Devasurendra AM, Kirchhoff JR. Selective determination of L-DOPA at a graphene oxide/yttrium oxide modified glassy carbon electrode. Electrochimica Acta. 2019;**301**:192-199. DOI: 10.1016/j.electacta.2019.01.098

[24] Hassanvand Z, Jalali F. Simultaneous determination of L-DOPA, l-tyrosine and uric acid by cysteic acid-modified glassy carbon electrode. Materials Science and Engineering C. 2019;**98**:496-502. DOI: 10.1016/j.msec.2018.12.131

[25] Prabhu P, Babu RS, Narayanan SS. Amperometric determination of L-dopa by nickel hexacyanoferrate film modified gold nanoparticle graphite composite electrode. Sensors and Actuators B: Chemical. 2011;**156**(2):606-614. DOI: 10.1016/j.snb.2011.02.006

[26] Sufredini HB, Pedrosa VA, Codognoto L, Machado SAS, Filho RCR, Avaca LA. Enhanced electrochemical response of boron-doped diamond electrodes brought on by a cathodic

surface pre-treatment. Electrochimica Acta. 2013;**49**:4021-4026

[27] Naushad M, Gupta VK, Wabaidur SM, AlothmanZA.Simultaneousdetermination of benserazide and levodopa in pharmaceutical tablet, human serum and urine sample by differential pulse voltammetry using modified glassy carbon electrode. International Journal of Electrochemical Science. 2013;**8**:297-311

[28] Kul D, Brett CMA. Electrochemical investigation and determination of levodopa on poly(Nile blue-a)/ multiwalled carbon nanotube modified glassy carbon electrodes. Electroanalysis. 2014;**26**:1320-1325. DOI: 10.1002/elan.201400071

[29] Zhou YZ, Alany RG, Chuang V, Wen J. Studies of the rate constant of L-DOPA oxidation and decarboxylation by HPLC. Chromatographia. 2012;**75**:597-606. DOI: 10.1007/ s10337-012-2229-1

[30] Ensafi AA, Bahmari H, Rezaei B, Maleh KH. Application of ionic liquid–TiO_2 nanoparticle modified carbon paste electrode for the voltammetric determination of benserazide in biological samples. Materials Science and Engineering. 2013;**33**(2):831-835. DOI: 10.1016/j.msec.2012.11.008

[31] Fouladgar M. A high sensitive square wave voltammetric sensor based on ZnO nanoparticle ionic liquid paste electrode for determination of benserazide in biological samples. Measurement. 2016:141-147. DOI: 10.1016/j.measurement.2016.02.057

[32] Rabinca AA et al. Voltammetric method for simultaneous determination of L-Dopa and benserazide. Current Analytical Chemistry. 2017;**13**:218-224. DOI: 10.2174/15734110126661606011 61703

[33] Rahmanifar E, Yoosefian M, Maleh KH. Application of CdO/ SWCNTs nanocomposite ionic liquids carbon paste electrode as a voltammetric sensor for determination of benserazide. Current Analytical Chemistry. 2017;**13**:46-51. DOI: 10.2174/1573411012 666160601145809

Chapter 6

Hybrid Graphene Nanocomposites: Thermal Interface Materials and Functional Energy Materials

Alexander S. Dmitriev

Abstract

Most existing materials may not satisfy all the fundamental requirements of modern civilization. This chapter summarizes the latest advances in the study of hybrid graphene nanocomposites and their application as thermal interface materials and some functional energy materials, in particular, for thermal management of energy and electronic devices. The main properties of hybrid graphene nanocomposites are described. The main attention is paid to the thermal properties of such materials, in particular, thermal conductivity and the possibilities of its growth due to various changes in the morphology and other properties of nanocomposites. The technology of obtaining a new nanocomposite based on mesoscopic microspheres, polymers, and graphene flakes is considered.

Keywords: nanocomposite, graphene, graphene nanoflake, thermal interface materials, thermal conductivity, thermal management

1. Introduction

The development of modern and promising materials is associated both with the need to study their fundamental properties and with their use in industrial applications, biology, medicine, and ecology [1–11]. Many of the existing materials may not satisfy all the fundamental requirements of modern civilization. This understanding prompted researchers to develop hybrid materials that may exhibit superior properties to those of the individual components. A special role in the development of new materials today is played by composites and hybrid composites. If in the base material (matrix) there are inclusions (fillers) of a certain size (microscopic, mesoscopic, or nanoscale), then the composite is classified as microcomposites, mesocomposites or nanocomposites [4, 8, 12–14]. Hybrid nanocomposites are composites in which, in addition to the base material (matrix) and nanofiller, other components of various sizes (microscopic, mesoscopic, or nanoscale) can be present. Hybrid graphene nanocomposites are composites that necessarily include carbon nanocomponents, for example, graphene in various modifications (nanoplates, nanoflakes, etc.). Hybrid graphene nanocomposites have attracted much attention recently because of their unique structure and remarkable mechanical, electrical, and thermal properties [15–20]. Recently, the development of hybrid nanocomposites has been growing at a rapid pace due to the numerous needs in energy and electronics, construction, aviation and space technology, biology and medicine.

Currently, there are various types of hybrid composites and nanocomposites, a description of the technology and properties of which are available in the scientific literature [9, 17, 21–31]. At the same time, the class of composites associated with nanocarbon components, namely graphene and its derivatives, has the most interesting and important properties. This does not mean that other classes of hybrid composites are less important, but there are hybrid graphene composites that demonstrate the most unexpected properties and prospects of use today.

For various applications, functional energetic materials must possess not only the necessary internal parameters, but also have a special functional surface (wettability, surface physical and chemical activity, etc.). The important areas of using functional energy materials based on graphene hybrid composites may include the following: superhydrophobic surfaces for anticorrosive protection of materials, surfaces for anti-icing, and self-cleaning surfaces [21]. Such materials are also used in boiling and condensation processes. The surface wettability plays a key role in boiling heat transfer, where the heat transfer coefficient is enhanced due to the rapid departure of nucleation bubbles from the surface with a small superheat. Condensation is a common process in industry, such as in the production of electricity, heat exchangers, and so on. If the surface energy of functional materials is high enough, film condensation actively occurs on it. Such materials include metals on which a decrease in the heat transfer coefficient is usually observed due to the presence of a liquid film that has low thermal conductivity. Coating a metal surface with a superhydrophobic material leads to dropwise condensation, which allows drops of liquid to easily break off the surface and be carried away into the external stream. It is the same properties that many graphene and hybrid nanocomposites have. Boiling heat transfer is often used in heat transfer devices, such as a heat exchanger, boiler, for cooling electronic and optoelectronic devices, solar thermal and photovoltaic energy, etc. due to its high heat transfer efficiency. As shown by numerous studies, the heat transfer efficiency on a superhydrophobic surface is much higher than on a smooth surface [18, 21]. The study of graphene nanocomposites as superhydrophobic materials for applications in power engineering, electronics cooling, and other applications is still insufficient. There are several publications devoted to this problem [20, 27]. However, practical data are not available for graphene nanocomposites. Therefore, it is necessary to study the heat transfer during boiling in various modes (nucleate and bubble growth, transition, and film), as well as critical heat fluxes on superhydrophobic surfaces of graphene nanocomposites. The topic of studying graphene nanocomposites as heat transfer, evaporation, boiling, and condensation surfaces is an important and relative new direction, which should be given great attention in the future [20–27].

2. Graphene nanocomposites (GNC): general properties

The hybrid material scheme is shown in **Figure 1**, which also shows the class of hybrid graphene composites, which will be described below. Published reviews and scientific articles contain very detailed information on the results of studies of such composites in recent years [14–18]. In **Figure 1**, the region of hybrid graphene nanocomposites (HGNC) is highlighted.

This chapter is devoted to this area of composites research. The term “graphene-related materials” is used to refer to graphene-related materials that have different names in scientific literature [1, 3–5, 8]. These include graphene oxides (GO), reduced graphene oxides (rGO), graphene foam (GF), graphene nanoribbons (GNR), graphene nanoplates (GNP), graphene sheets (GS), graphene nanoflakes (GNF), graphene film (Gfi), etc. The term "graphene composites" means that graphene components are placed in an organic or inorganic matrix. Moreover, the morphology of graphene components, their number, volume fraction, etc. can be completely different [24–27].

Figure 2 gives a simple idea of standard and hybrid composites. Here, you can see that usually standard composites consist of a base (mother) matrix and one type of filler. In hybrid nanocomposites, not only more than one filler, but in most situations there should be special morphology and architecture.

Hybrid graphene nanocomposites are currently used in various applications, as shown in **Figure 3**. Among these applications are functional energy materials for energy conservation, lithium-ion batteries, supercapacitors, fuel cells, hydrogen storage, systems for conversion solar radiation into steam and systems for producing clean and desalted water during solar heating, as well as many others [6, 18, 20–23]. Technologies and synthesis of graphene components, including graphene plates and flakes, are given in [15, 17, 19]. Currently, there are several methods for producing such graphene components on an industrial scale [19].

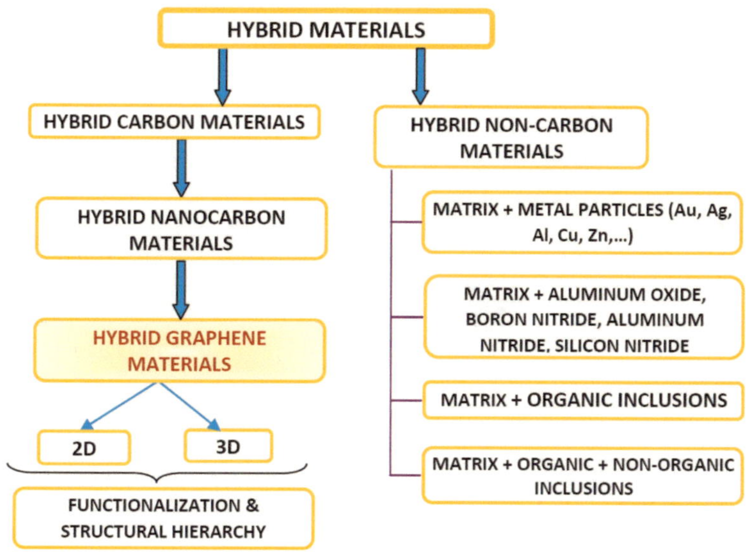

Figure 1.
Scheme hybrid graphene materials.

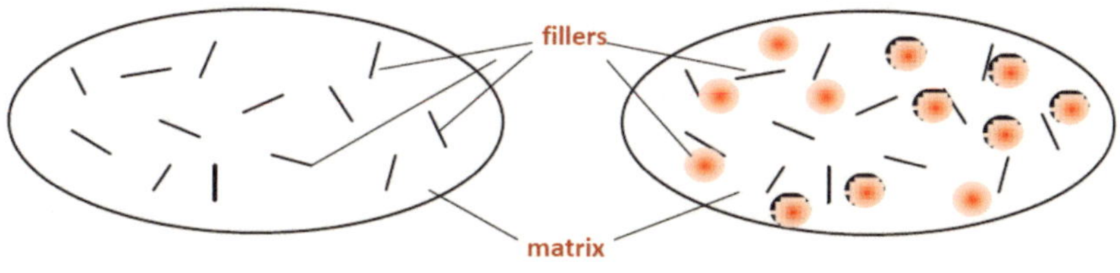

Figure 2.
Conventional (left) and hybrid (right) composites.

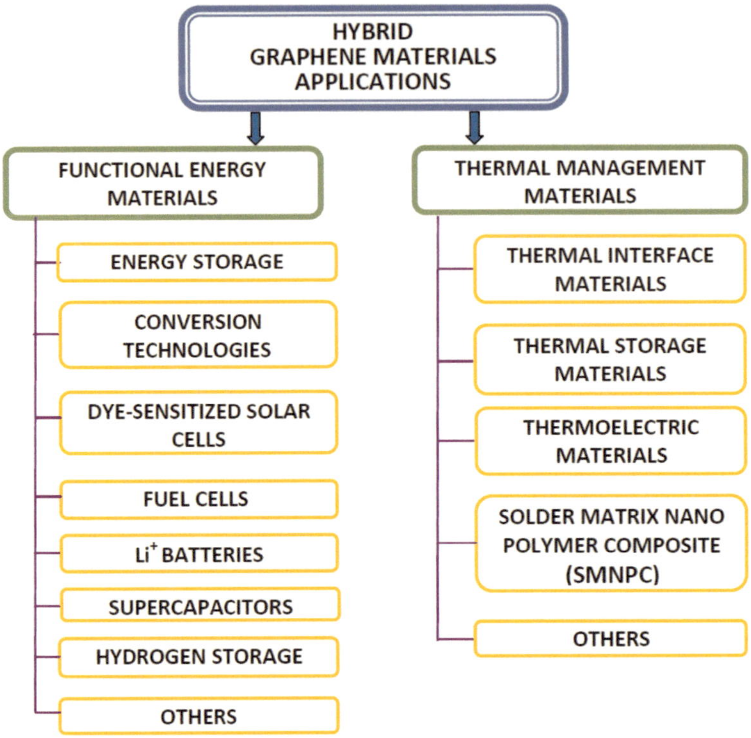

Figure 3.
Scheme for hybrid graphene materials application.

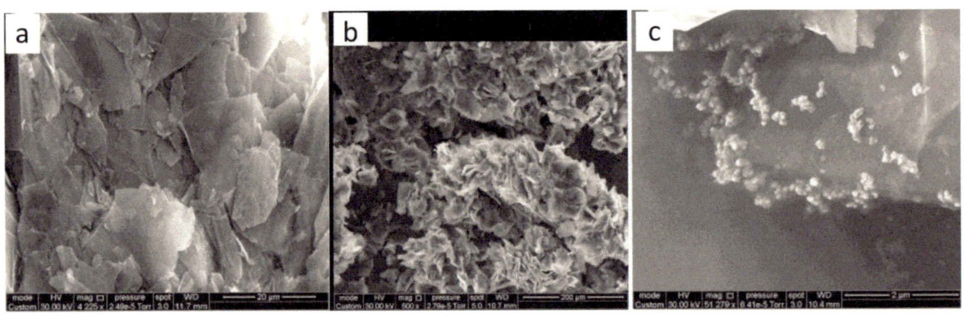

Figure 4.
Electron microscopy of hybrid graphene composites: a - graphene flakes, b - graphene flakes on polymer droplets, c - gold nanoparticles on the surface of graphene flakes.

Graphene hybrid materials have been fabricated by the cross-linking of graphene or graphene oxide (GO) through various kinds of inorganic or organic species, i.e., inorganic nanoparticles, polymers, multifunctional organic molecules, and metal ions/complexes [4, 7, 8]. More recently, a review has been published that provides many important data on graphene and hybrid nanocomposites [32]. The table contains the main research results in recent years on graphene nanocomposites [20, 28, 33–52].

As an example, **Figure 4** shows the main components when creating graphene hybrid nanocomposites. Graphene flakes with lateral (longitudinal) sizes of 10–30 μm, a thickness of 3–5 nm, obtained by liquid-phase exfoliation of graphite in pure water (**Figure 4a**); graphene flakes on polymer droplets (prior to compression and thermal treatment and obtaining a polymer matrix (**Figure 4b**)). Gold nanoparticles of 4–6 nm in size on the surface of graphene flakes produce hybrid graphene nanocomposites (**Figure 4c**).

The graphene-based inorganic or organic hybrid materials have been extensively investigated in various applications: as thermal interface materials, functional energy materials, energy storage, and/or conversion-related fields [32, 33]. In particular, the following were investigated:

- effect of filler loading level,
- fillers of various dimensions: one-dimensional fillers (1D fillers with high-aspect-ratio), platelet-like fillers, spherical and quasispherical fillers,
- surface treatment,
- filler orientation,
- filler agglomerates,
- formation of continuous filler network (thermal percolation effect),
- double percolation (thermal conductivity and electrical conductivity percolation effects),
- functionalization of fillers, and
- size effects.

Fillers	Category	Thermal conductivities W/(m K)
Aluminum (Al)	Metal	234
Copper (Cu)	Metal	386
Gold (Au)	Metal	315
Silver (Ag)	Metal	427
Carbon nanotube (CNT)	Carbon based	1000–4000
Carbon fiber	Carbon based	300–1000
Graphite	Carbon based	100–400
Graphene	Carbon based	2000–6000
Diamond	Carbon based	9000
β-Silicon nitride (β-Si_3N_4)	Ceramics	103–200
Hexagonal boron nitride (h-BN)	Ceramics	185–300
Aluminum nitride (AlN)	Ceramics	100–300
Diamond like	Ceramics	1000
β-Silicon carbide (β-SiC)	Ceramics	120
α-Alumina (α-Al_2O_3)	Ceramics	30
Beryllium oxide (BeO)	Ceramics	270
ZnO	Ceramics	21
SiO_2	Ceramics	1

Table 1.
Thermal conductivities of common types of fillers.

Filler # 1	Filler # 2	Filler # 3
h-BN	Different sized h-BN	—
h-BN	Carbon fiber	—
h-BN	Carbon nanotubes (CNTs)	—
Carbon nanotube (CNT)	Carbon based	—
Hollow glass microspheres	Nitride nanoparticles	—
Aluminum nitride (AlN)	Carbon nanotubes (CNTs)	—
SiC (nanosized)	Carbon nanotubes (CNTs)	—
Graphite	Carbon fiber	—
Graphene	BN	—
Laminated h-BN	SiC microparticles	—
Graphite nanoplatelets	Ceramics	—
Graphene nanoflakes	Metallic microspheres	—
Graphene nanoflakes	Metallic microspheres	Au, Ag, SiO_2 nanoparticles
Graphene nanoflakes	Carbon nanotubes (CNTs)	Metallic microspheres
Graphene nanoflakes	Graphite	

Table 2.
Effects of hybrid fillers.

Table 1 presents the main fillers of polymer composites when the composites contain only one type of filler [33].

Table 2 shows the hybrid nanocomposites options in which there are several types of fillers [33].

3. Thermal conductivity of hybrid nanocomposites and graphene nanocomposites

The morphology of graphene in a polymer matrix significantly affects the thermal conductivity of composites [33]. It is usually customary to separate morphology into two types: random dispersion of graphene in a polymer matrix and regular [20, 32]. Random dispersion refers to the addition of graphene to a matrix, which is performed by a simple method, such as agitation, sonication, and blending. In addition, there is usually no special method used to control the orientation of graphene in the matrix. The second type of morphology is graphene with a specific orientation in the polymer matrix (regular structure). This refers to unusual graphene structures in a polymer matrix, including orientation, three-dimensional structure (3D) and separate structure, etc. Graphene with a random orientation in the polymer matrix can be manufactured by many methods, such as solution mixing, melt mixing, and in situ polymerization, etc. [32].

Some thermal properties of graphene composites with random orientation are given in **Table 3** [20, 32, 33, 40, 42]. The special orientation of graphene can significantly affect the properties of graphene nanocomposites. In the case of hybrid nanocomposites, the situation can be even more complicated: the mutual orientation of several types of fillers can affect the final properties of graphene materials.

The specific orientations of graphene give composites special properties. The thermal properties of recent studies of graphene-polymer composites are shown in

Nanocomposites	Filler loading	Thermal conductivities W/(m K)	Thermal conductivities enhancement per wt%
Epoxy + GNS-Py-PGMA	3.8 wt%	1.91	225%
Epoxy + f-Gfs	10 wt%	1.53	66.5%
Epoxy + GNP + C750	5 wt%	0.45	23.8%
Epoxy + DGEBA + f-GO	4.6 wt%	0.72	52.3%
PVDF + GS + Al_2O_3	40 wt%	0.586	4.8%
Epoxy + GNP + Al_2O_3	12 wt%	1.49	56.4%
PBT + GNP	20 wt%	1.98	61%
PPS + GNP	37.8 vol%	4.14	49%
PI + SiCNW-GS	7 wt%	0.58	21%
SR + GNP	0.72 wt%	0.3	69.4%
PA6 + Gr-GO	10 wt%	2.14	56.9%
Epoxy + GNP	25 vol%	2.67	49.4%
PVDF + FGS + ND	45 wt%	0.66	3.9%
Epoxy + ApPOSS-Gr	0.5 wt%	0.35	115.8%
PU + IL-Gr	0.61 wt%	0.3	55.9%
PA + TCA-rGO	5 wt%	5.1	357.8%
BE + Gr	2.5 wt%	0.54	73.7%
Silicone + GNP	16 wt%	2.6	49.7%

Table 3.
Graphene composites with random orientation.

Table 4 [32, 33, 40]. Most studies have shown that an increase in the volume concentration of the nanocomponent practically does not affect the crystal structure of materials.

Traditional polymer composites cannot meet the requirement of achieving higher thermal conductivity at relatively low filler loading. Regular orientation of fillers is important in the thermal conductivity of nanocomposites. In particular, when controlling the orientation of silicon carbide nanowires (with a very low filler loading of about 5 wt%) in epoxy resins, the thermal conductivity of the composite in the plane reached 10.1 W/m K. On the other hand, with random orientation of SiCNW nanowires in epoxy resin, the thermal conductivity was only 1.78 W/m K, and for an epoxide/silicon carbide composite in the form of nanoparticles, it was 0.30 W/m K [53]. Theoretical models of the thermal conductivity of such composites show that the correct orientation and high aspect ratio for nanowires contribute to the formation of heat transfer networks in composites, leading to the effect of thermal percolation.

Currently, the main types of thermal interface materials used have a matrix of organic compounds, for example polymers, which are filled with inorganic particles of high concentration, such as ceramic or metal. Recently, graphene nanoplates (GNPs) or nanoflakes (GNFs) [27], usually having several graphene layers, have been actively used as fillers. Due to the special morphology of the layers, as well as

Nanocomposites	Filler loading (Gr wt%)	Thermal conductivities W/(m K)	Thermal conductivities enhancement per wt%
PDMS + vertically aligned Gr film	92.3 wt%	614.85	3329%
Epoxy + multilayer Gr	11.8 wt%	33.54	1413%
Epoxy + nanofibrillated cellulose	1 wt%	12.6	910%
PVDF + large-area rGO	27.2 wt%	19.5	324%
Oriented GNF + PVDF	36.8 wt%	10	113%
Epoxy + GNS-Fe_3O_4	1.74 wt%	0.6	80%
Polyethylene + GNP	10 wt%	1.84	45.7%
Polypropylene+ GNP	10 vol%	1.53	59.5%
PVA+ GNP	10 wt%	1.43	58%
PVDF + GNP	10 wt%	1.47	67.3%
Epoxy + Gr + SWCNT	7.5% wt%	1.75	—
Epoxy + Gr + MWCNT	20 wt%	6.31	—
Epoxy + Ag flakes + CNT (functionalized)	35.5 wt%	160	—
Epoxy + hBN + SiCNW	95 wt%	21.7	—
Epoxy + hBNNT	30 wt%	2.77	—
PVDF + AgNW	25 vol%	1.61	1050%
Epoxy + CoNW	0.12 vol%	2.59	700
PVDF + Cu_2ONW	30 wt%	0.32	170
Epoxy + SiO_2 (coated AgNW)	4 vol%	1.03	415
Epoxy + SiNNW	60 wt%	9.2	4281
Epoxy + BNNT	30 wt%	2.77	1360

Table 4.
Thermal properties of graphene-polymer composites.

hybridization of the chemical bonds of the carbon lattice, such plates or flakes have a very high thermal conductivity [54, 55], which makes it possible to control the thermal conductivity of organic matrices even at very low concentrations. It was noted that an increase in the lateral size of graphene nanoflakes and the number of layers leads to an increase in their thermal conductivity, and, consequently, to the thermal conductivity of the nanocomposite. Apparently, there are an optimal number of graphene layers, which ensures low phonon scattering and leads to high thermal conductivity of nanoflakes with a small number of layers (less than 10) [56, 57].

The production of nanofilled GNP resins is crucial for the final properties of these materials, since they are highly influenced by dispersion, stratification and orientation levels, as well as the final morphology (size, waviness, imperfection) of GNP. For the creation of heat-conducting materials, dispersion, orientation, and the degree of delamination are important [58]; the presence of defects can reduce thermal conductivity; poor interaction between the matrix and nanofillers can lead to a decrease in mechanical strength and deterioration of thermal properties, etc.

Thus, the most important task is a thorough analysis and selection of the production route and initial characteristics of GNP (functionalization and morphology) [59].

Oriented 3D-BN network composites in epoxy were made by combining self-assembly and infiltration technology using ice patterns [60]. It was shown that the resulting composites have good thermal conductivity −4.42 W/m K and a lower coefficient of thermal expansion. The authors attribute the improvement in thermal conductivity to the following factors: well-aligned BN plates in the direction of higher thermal conductivity, since the latter is anisotropic, as well as a decrease in the boundary thermal resistance. In addition to this, apparently, there is a three-dimensional network of BN plates in epoxy resin. In particular, there is a difference in the thermal interface resistance of oriented and random composites (4.0×10^{-7} and 5.6×10^{-7} m^2 K/W, respectively). Thus, the importance of orientation on the behavior of the thermal conductivity of nanocomposites is obvious. The dependence of the thermal conductivity of the nanocomposite on random and oriented plates and on their volume concentration is shown in **Figure 5** (left). **Figure 5** (right) shows the temperature dependence of the thermal conductivity of such nanocomposites. It should be noted that at a certain temperature, a drop in heat conduction is observed, which is an important factor when using such materials for thermal management.

As previously established, one of the most important factors associated with the thermal conductivity of nanocomposites filled with graphene nanoparticles (GNP) is the size of the GNP, that is, their lateral size and thickness. A relationship was established between the thermal conductivity of polymer composites and the size of GNP fillers in polymer composites.

An increase in the thickness of graphene plates and their transverse size leads to a decrease in the boundary thermal resistance between them and the matrix. In particular, for a volume fraction of GNP, 20 wt% volumetric and in-plane thermal conductivity was 1.8 and 7.3 W/m K, respectively (thermal conductivity polymer matrix 0.24 W/m K) [60].

The use of inorganic matrices is proposed in [61]. Hybrid paper with graphene/SiC (graphene hybrid paper, GHP) was developed by an easy and easily scalable filtration method followed by rapid heat treatment to grow SiC nanorods in situ between graphene sheets based on the carbothermic reduction reaction. GHP demonstrates a characteristic structure consisting of a hierarchical architecture of graphene/SiC nanorods, which leads to an increase in thermal conductivity in the plane (10.9 W/m K) by 60% compared to graphene paper. It is interesting to note

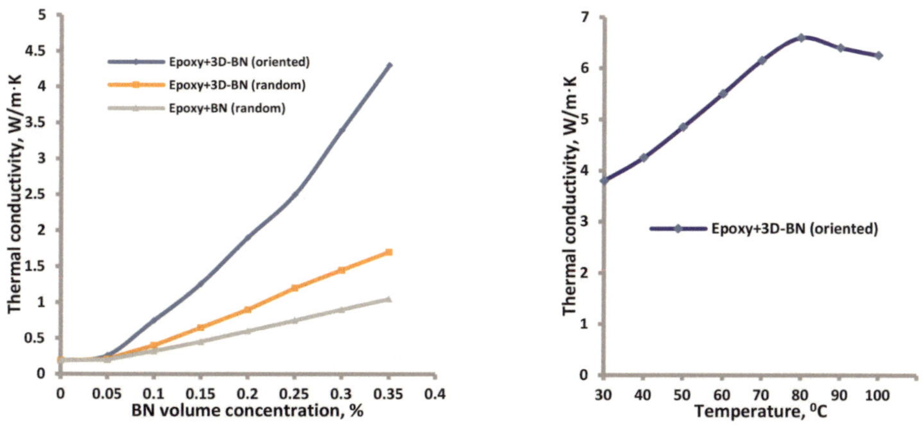

Figure 5.
Thermal conductivity BN nanocomposites [60].

that graphene-based paper, if used as TIM, significantly reduces thermal conductivity through the plane when compressive force is applied. The presence of the C—Si covalent bond leads to an increase in thermal conductivity to 17.6 W/m K in the presence of compressive force. Eliminating the aging problem of conventional polymer-based TIMs, GHP with its characteristic inorganic structure has great potential for use as highly effective TIMs with good thermal and chemical stability.

Three modern TIMs based on graphene are compared, including dispersed graphene/polymers, graphene framework/polymers, and inorganic monoliths based on graphene [28, 62]. Their advantages and limitations are discussed in terms of application. In addition, potential strategies and future research directions in the field of the development of highly efficient graphene-based TIMs are discussed.

Recently, a unique design of hybrid nanocomposites was used: microscale flakes Ag and multiwalled carbon nanotubes (MWNTs), which were decorated with Ag nanoparticles (nAgMWNT), were placed in a polymer matrix [63]. It was shown that even a small volume fraction of nAgMWNT (2.3 vol%) in epoxy matrix leads to the creation of effective phonons transfer paths between Ag flakes (35.8 vol%) (thermal conductivity ~160 W/m K). The successful dissipation of computer CPU heat using nAgMWNT-Ag-flake-epoxy TIM demonstrates the superior ability to cool electronics.

In addition to graphene hybrid nanocomposites based on epoxy resin, other polymers are also used. In particular, polymer composites consisting of graphene foam (GF), graphene sheets (GSs), and flexible polydimethylsiloxane (PDMS) were made, and their thermal properties were studied [46]. Due to the unique interconnected GF structure, the thermal conductivity of the GF/PDMS composite reaches 0.56 W/m K, which is approximately 300% higher than that of pure PDMS and 20% higher than that of GS/PDMS composite with the same graphene load of 0.7 wt%. **Figure 6** shows the dependence of thermal conductivity on temperature and volume fraction.

Coefficient of thermal expansion is (80–137) $\times$ 10^{-6}/K in the range of 25–150°C, which is significantly lower than that of composites GS/PDMS and pure PDMS.

In addition, it also exhibits excellent thermal and dimensional stability. Later, these authors added a different amount of multilayer graphene flakes (MGF) to the composition of 0.2 vol.% GF/polydimethylsiloxane (PDMS). It was noted that in such a graphene hybrid nanocomposite, a synergistic effect between MGF and GF was achieved in improving the thermal conductivity of polymer composites. When

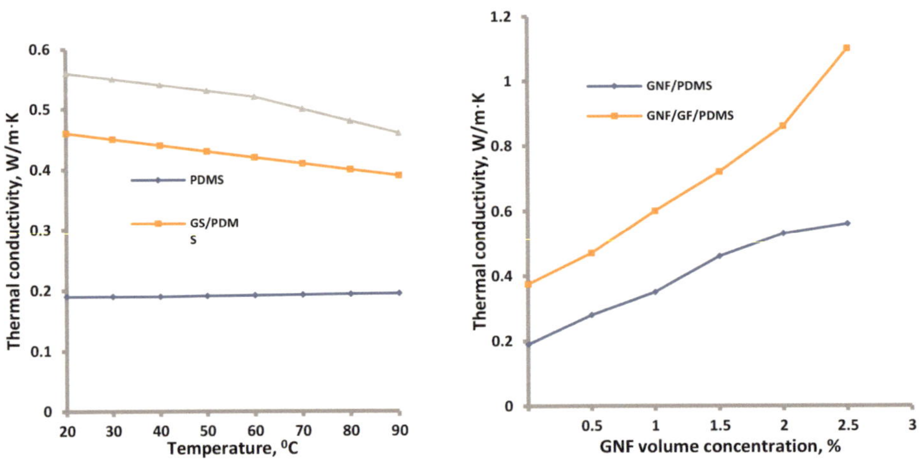

Figure 6.
Thermal conductivity on temperature (left) and volume fraction GNF (right).

the MGF content is 2.7 vol%, the thermal conductivity of the MGF/GF/PDMS composite reaches 1.08 W/m K, which is 80, 184 and 440% higher than that of the MGF/PDMS composites 2.7 vol%, GF/PDMS and pure PDMS, respectively. The MGF/GF/PDMS composite also exhibits excellent heat resistance. Adding MGF and GF slightly reduces elongation at break, but significantly increases Young's modulus and tensile strength of composites compared to pure PDMS. The good performance of the MGF/GF/PDMS composite makes it a good TIM for possible applications in thermal control of electronics.

An important area of the use of graphene nanocomposites is functional energy materials, among which a special role is played by phase change materials (PCM) for thermal energy storage (TES) used in a wide range of applications, including control of thermal electronic devices and thermal storage of solar energy. It is proved that this is an effective method for thermal power plants due to its high heat capacity and small temperature changes [41]. However, most PCMs have low thermal conductivity, which reduces the efficiency of the thermal storage device. In order to increase the PCM heat transfer ability and prevent the leakage of molten PCM, the TES system requires a heat exchange amplifier and a container [2–8]. Researchers used metal foam, additives, or fins to increase PCM thermal conductivity. However, these amplifiers add significant weight to the TES, and some of them are incompatible with PCM. This is the reason for the search for hybrid nanocomposites with a wide range of properties. First of all, in recent years, attention has been paid to hybrid graphene materials [21–23].

The new three-dimensional hierarchical graphene foam material (HGF) was obtained by filling the pores of GF with hollow networks of graphenes [38]. A hybrid nanocomposite based on paraffin (PW, matrix) and HGF showed a thermal conductivity of 744% higher than that of pure PW. To improve the properties of such materials, further hybrid graphene materials with the addition of multiscale fillers will undoubtedly be needed in the future [64].

It was found that the increase in thermal conductivity of composites with thicker graphene fillers (GNF) from several layers is greater than that of composites with thinner fillers with the same loading fraction [48] (**Figure 7**). The deviation found from the linear dependence of thermal conductivity on the volume concentration of

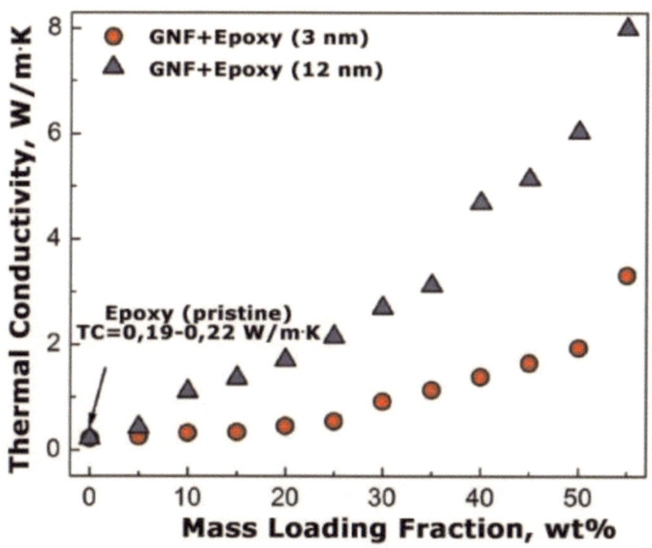

Figure 7.
Thermal conductivity on loading fraction and thickness GNF [48].

graphene nanoflakes indicates the beginning of thermal percolation in graphene composites.

The increase in thermal conductivity of hybrid nanocomposites can be achieved in various ways: the use of fillers with high thermal conductivity, the use of hybrid fillers, the creation of a new architecture of nanocomposites, including the effects of thermal percolation, etc. In particular, the last effect can be demonstrated: the heat flow through the effect of thermal percolation is shown in **Figure 8**.

In [65], a new tunable HGNC containing a combination of two different fillers based on carbon, graphene nanoplates (GNP) and graphite was proposed. By adjusting the concentration ratio of GNP: graphite and the total concentration of fillers, the authors were able to fine-tune the thermal conductivity and workability of HGNC. The following filler parameters were used: the average length of the filler particles was determined by measuring 100 particles for each filler material and the sizes of the different fillers, when imbedded in the epoxy matrix, are 19 ± 3 and 27 ± 4 μm for the GNP and the graphite, respectively. **Figure 9** shows the dependences of the thermal conductivity of nanocomposites on the volume fraction of graphite. It is clearly seen that such dependencies are linear. With an increase in the

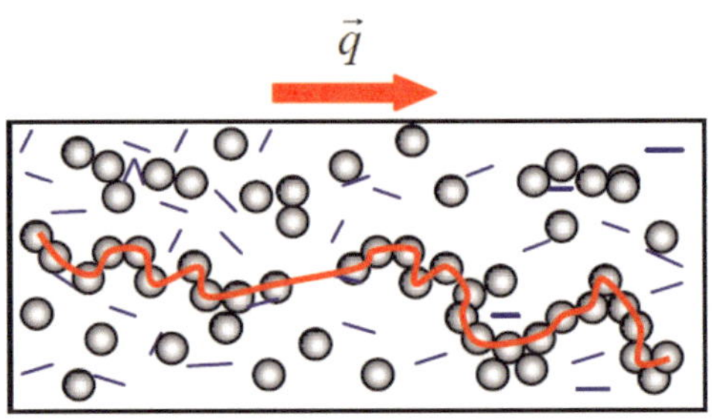

Figure 8.
Heat flow through the effect of thermal percolation.

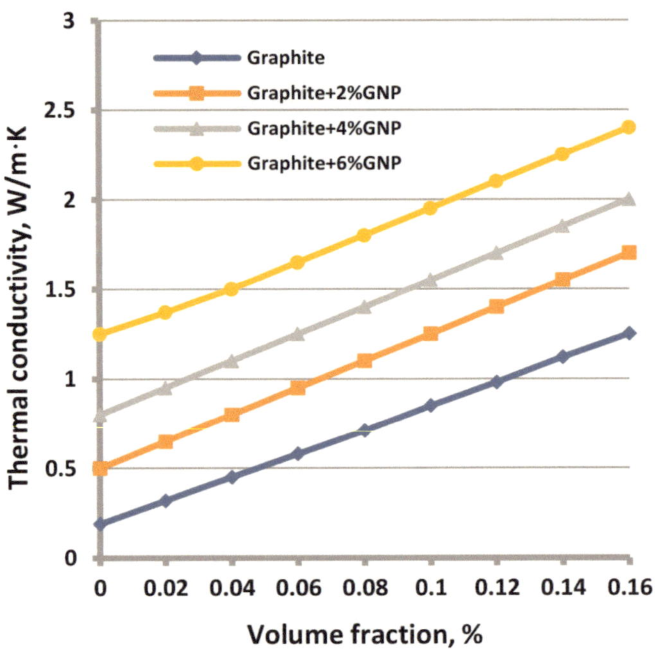

Figure 9.
Thermal conductivity of hybrid composites (T = 25°C) as a function of the graphite volume fraction [65].

volume fraction of graphene nanoplates, the overall thermal conductivity increases, which corresponds to the general trend of the effect of graphene nanoplates on heat transfer.

To optimize the design of composite materials, the authors also studied the effect of viscosity of the starting material and built a phase diagram of the concentration-thermal-conductivity viscosity. Thus, this ensures not only the selection of the thermal conductivity of the desired nanocomposite, for example, for TIM, but also the manufacturability of the material itself.

4. New hybrid grapheme nanocomposites

4.1 Hybrid graphene nanocomposites: preparation and properties

When creating hybrid graphene nanocomposites, fillers of approximately the same size, for example, nanoscale, are usually used. Several studies have attempted to introduce microsized components along with nanoscale components. However, for some reason, it was not possible to achieve a noticeable increase in the thermal conductivity of such composites [66–69]. In the framework of this study, an attempt was made to combine the remarkable properties of graphene flakes with metal microparticles, which in themselves have high thermal conductivity. On the other hand, it was assumed that the use of equally sized (monodisperse) metal microspheres would help to make a regular controlled structure of the composite. We also note that the use of metal particles makes it difficult to use a similar material for thermal interfaces, since the latter must be nonconductive. This problem was solved by the fact that the microspheres, for example, tin, were partially oxidized on the surface, which led to a significant potential barrier to the flow of electric current (noticeable electrical boundary resistance). At the same time, as the research results showed, the boundary thermal resistance of the metal + oxide film boundary was always less than the boundary thermal resistance at the boundary of the microsphere with the polymer.

Hybrid composite functional energy materials require new approaches, the selection of new components, and their multiscale properties. New architectures have been developed for hybrid composites based on metal microspheres, polymers (e.g., epoxy), and graphene nanoflakes (GNF). According to the technology described in [66, 67], graphene nanoflakes with lateral sizes of 5–20 μm and 3–10 layers of graphene were obtained.

Electron microscopy of graphene nanoflakes is shown in **Figure 10**. It is clearly seen that graphene flakes can be stacked (by filtration) in regular structures. After that, the microspheres (100–150 μm) from Sn, Pb, and Er3Ni (production of the Moscow Power Engineering Institute) were placed in a petri dish and a nanocolloid solution of graphene flakes was poured into it (**Figure 11**). **Figure 6** shows metal microspheres from Sn coated with graphene nanoflakes. It is clearly seen that the resulting structure is very regular, and nanoflakes can adhere well to the microspheres. It should be noted that epoxy resin and polyurethane interact well with graphene flakes and metal microspheres, with significant adhesion. This allows you to create dense structures you need morphology. More detailed description of materials and methods is given in our publications [66, 67]. For the created hybrid graphene nanocomposites, the mechanical, electrical, and thermal properties were investigated. Below are the results on the thermophysical characteristics of such materials.

Figure 12 shows electronic images of graphene flakes in various polymer matrices.

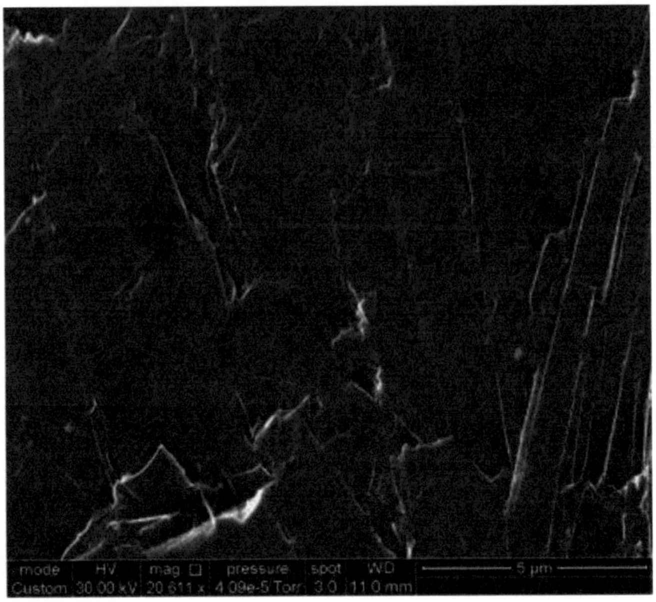

Figure 10.
Electron microscopy of graphene nanoflakes.

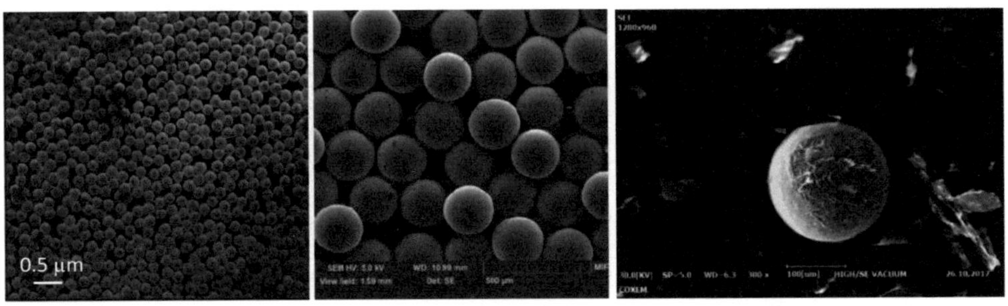

Figure 11.
Metal microspheres from Sn coated with graphene nanoflakes.

Figure 12.
The structure of graphene nanoflakes in a polymer matrix (left, polyurethane and right, epoxy resin).

Differential calorimeter and laser flash methods were used to study the thermal conductivity of the developed composites. The results are shown in **Figure 13**.

4.2 Thermal conductivity model of hybrid nanocomposites

Figure 14 shows two types of hybrid thermal interface materials with graphene flakes (dark inclusions), metal microspheres, and a polymer (gray field). On the left

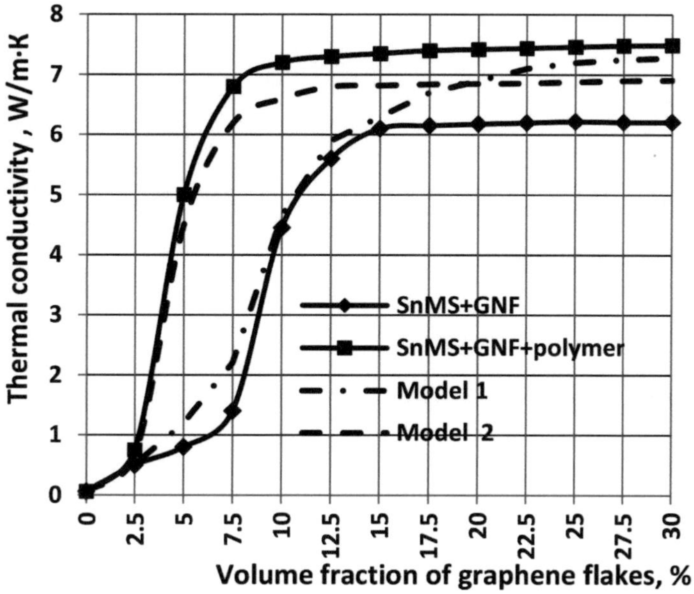

Figure 13.
Thermal conductivity as a function of volume fraction of graphene flakes.

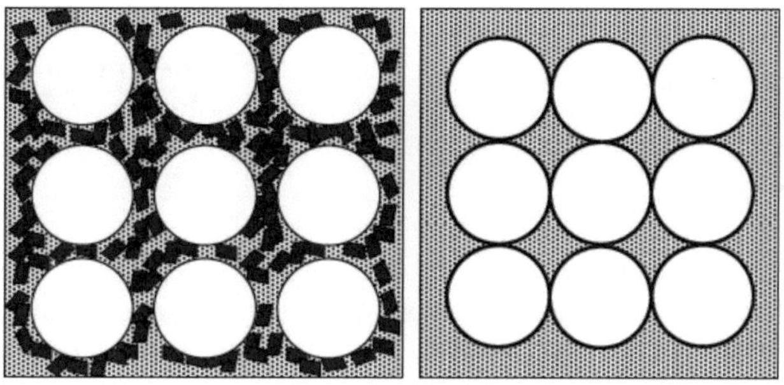

Figure 14.
Structure HGNC model.

are free-packed microspheres without polymer, on the right are dense packaging of microspheres, between which graphene flakes are located, the adhesive of the nanocomposite is polymer (graphene flakes are located on the surface of the microspheres).

The thermal conductivity of hybrid nanocomposites was calculated within the framework of a modified model of the effective medium, which leads to the following expression for the effective thermal conductivity of the thermal interface

$$\lambda_{TIM} = \frac{3\lambda_m + 2\phi\left\{\lambda^p_{GNF}\left[1 + \left(2R_K\lambda^p_{GNF}/L\right)\right]^{-1} - \lambda_m\right\}}{3 - \phi[1 - (2R_K\lambda_m/\delta)]} \quad (1)$$

where $\lambda_m \approx 0.21$ W/m K is the thermal conductivity of the matrix (polymer-epoxy resin), λ^p_{GNF} is the thermal conductivity along the graphene flakes plane (GNF in-plane thermal conductivity), ($\lambda^p_{GNF} \approx 1670$ W/m K), $R_K \approx 6 \times 10^{-8}$ m^2 K/W is the boundary thermal resistance (Kapitsa resistance) between the polymer and GNF (estimated), is the typical longitudinal size of grapheme flakes (in our case $L \approx 10$–20 μm), and $\delta \approx 3$–3.5 nm is the thickness of GNF.

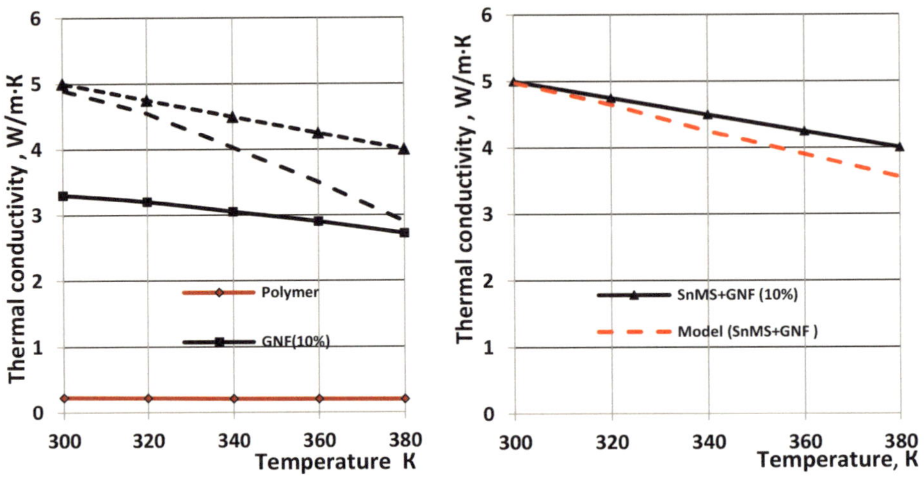

Figure 15.
Dependences of thermal conductivity for various composites.

The results of calculating the effective thermal conductivity by the above ratio are presented in **Figure 15**. It is clearly seen that with an increase in the volume fraction of graphene, the thermal conductivity monotonically increases. However, both for the microsphere + graphene composite and for the composite with the polymer, it saturates (for the case without polymer, approximately starting from 7.5%, for the case with polymer, from 12.5%). At the same time, it is easy to see that the model for freely mixed microspheres, GNF, and polymer leads to increasing thermal conductivity, while in the experiment, the latter is saturated. At the same time, with a GNF volume fraction of about 20%, the difference is more than 10%.

Thus, for such a system, the experiment shows a lower value of the heat transfer efficiency, which, apparently, is associated with a large value of the boundary heat resistance. For the case of the second type (dense hybrid nanocomposites), the situation is completely different. In this case, the calculation predicts a lower value of thermal conductivity at all volume concentrations of GNF. In our opinion, this fact speaks of a specific phenomenon of "thermal percolation," the mechanism of which is discussed in more detail below.

The temperature dependences of the effective thermal conductivity of hybrid nanocomposites were also calculated (**Figures 13** and **15**). Calculations showed that thermal conductivity with temperature decreases more than in experiment, and the difference between the calculated and experimental values at a temperature of 380 K is more than 33%. This is a very strong difference in thermal conductivities, which may also indicate the influence of specific mechanisms not taken into account in theoretical models.

4.3 "Thermal percolation" through the interface and modification of the ratios for effective thermal conductivity

Usually, when considering the boundary thermal resistance of Kapitsa, it is assumed that the thermal interface does not change with external exposure. On this basis, two main models of the boundary thermal resistance are formulated—the models of acoustic and diffusion impedance: in the first model, it is assumed that phonons scatter elastically (specularly) at the boundary; in the second, there is diffuse scattering (phonons forget the incidence conditions at the boundary); lose memory during scattering) and the probability of passing through the interface depends on the ratio of the densities of phonon states on each side. On the other hand, if an external action can change the interface, then one must take into account the behavior of the nonequilibrium phonon function when approaching the interface. It is generally believed that phonons are suitable for the interface in ballistic mode (without internal resistance of the interface), and all temperature drop is formed on the interface itself. We define the interface between the two materials 1 and 2; phonons that fall from body 1 and have a wave vector for the mode can either be reflected back to 1 or pass to body 2. We set the probability of passage from 1 to 2 equally, then the thermal boundary conductivity determines the heat flux from 1 to 2 for a given temperature difference in the form

$$G = \frac{1}{2(2\pi)^3} \sum_i \int_{\vec{k}} \Xi_{12}\left(\vec{k}, i\right) \left[\hbar^2 \omega_i^2\left(\vec{k}\right)/k_B T^2\right] \vec{v}_i\left(\vec{k}\right) \times \vec{n} \exp\left[\hbar\omega_i\left(\vec{k}\right)/k_B T\right] \left[\exp\left(\hbar\omega_i\left(\vec{k}\right)/k_B T\right) - 1\right]^{-2} d\vec{k} \qquad (2)$$

where $\omega_i\left(\vec{k}\right)$, $\vec{v}_i\left(\vec{k}\right)$ are the frequency and group velocity of phonons, respectively, with a wave vector $\vec{k}$ for i mode in medium 1; $\vec{n}$ is the unit normal wave vector to the surface between the bodies 1 and 2. If the dispersion relation for phonons $\omega_i\left(\vec{k}\right)$ is known, as well as the probability of propagation $\Xi_{12}\left(\vec{k}, i\right)$, then the integral can be calculated in explicit form. It is clear that the phonon spectrum on both sides of the boundary is known very approximately. Therefore, in the gray approximation, all quantities included in the expression (2) were calculated. Despite the low accuracy of this approximation, it was possible to calculate the basic quantities for heat transfer through the interface.

In the diffusion model, it is assumed that phonons falling from 1 to 2 forget their state and scatter with the same energy in another medium $\Xi_{21}(\omega') = 1 - \Xi_{12}(\omega')$. This means that the transmission coefficient is the same for similar states, i.e., passing from medium 2 to medium 1 takes place. We assume that the scattering is elastic, so that the phonon frequency does not change. In this situation, it is possible to record the probabilities of passage from medium 1 to medium 2 and vice versa in the form of transmission coefficients and. Moreover, the reflection probabilities are respectively equal. It is also believed that the probabilities of passage and reflection in the sum are equal to unity. If the temperatures on both sides of the interface are the same, then the phonon fluxes from medium 1 to 2 and vice versa are equal. If and are phonon flows incident on the interface from each side, then there is a relation $\Xi_{12}(\omega) j_1(\omega) + \Xi_{21}(\omega) j_2(\omega) = 1$ and $\Xi_{12}(\omega) = \left[1 + j_1(\omega)/j_2(\omega)\right]^{-1}$. The obtained relations for the phonon transmission coefficients through the interface, taking into account the mismatch in the limiting cases of the acoustic model (AMM) and diffusion model (DMM), solve the problem of determining the thermal contact resistance in the macroscopic approximation. Bearing in mind the above relations for the boundary thermal resistance, we obtain corrections to the effective thermal conductivity, which was used earlier.

The main idea is that when a hybrid nanocomposite is compressed between metal microspheres, an extremely thin layer of polymer with GNF appears, and the concentration of the latter exceeds the threshold of "thermal percolation"; i.e., the heat flux almost passes through graphene flakes. In this case, the thermal conductivity should noticeably increase compared with the case of a nanocomposite with thick polymer interlayers. Indeed, calculations of the percolation conditions lead to the conclusion that with a GNF volume fraction inside the polymer layer exceeding

48%, a regime of direct phonon propagation along and across graphene flakes arises. In this case, an additional contribution to the heat flux appears, which is shown in **Figure 15**. It is clearly seen that compared with **Figure 13** where the difference at a temperature of 380 K was more than 33%, taking into account thermal percolation, the difference at this temperature is only 11%. This fact indicates the importance of considering the heat transfer in hybrid nanocomposites to take into account the inhomogeneous distribution of GNF inside the polymer interlayer. Thus, the analysis of heat transfer in hybrid nanocomposites consisting of graphene flakes, monodisperse metal microspheres and polymers showed a very complex behavior of heat transfer. The heat transfer mechanisms calculated on the basis of the modified theory of effective thermal conductivity showed that the main contribution to the heat flux comes from the boundary thermal resistance between microspheres, graphene flakes, and thin layers of polymers. By comparison of the obtained results with the experimental data published by us earlier, it is shown that the proposed model correctly describes the thermal conductivity of the hybrid nanocomposites; however, at high temperatures, there are very significant differences from the experimental data, which can be explained by a decrease in the boundary thermal resistance at the interface due to "thermal percolation." In our opinion, more thorough studies of these effects are necessary in order to identify the nature of heat transfer in complex hybrid nanocomposites based on graphene flakes. Only this will allow confidently creating and calculating thermal interface materials for thermal stabilization and cooling of electronics and energy devices.

Prospects for the use of hybrid graphene nanocomposites in the future are very interesting and may bring many unexpected results.

Acknowledgements

This work was supported by the Russian Science Foundation (project No. 17-19-01757).

Author details

Alexander S. Dmitriev
Nano Thermo Lab and Low Temperature Department, National Research University "Moscow Power Engineering Institute", Moscow, Russia

*Address all correspondence to: asdmitriev@mail.ru

References

[1] Kickelbick G. Hybrid materials–Past, present and future. Hybrid Materials. 2014;**1**:39-51. DOI: 10.2478/hyma-2014-0001

[2] Papageorgiou D, Kinloch I, Young J. Graphene/elastomer nanocomposites. Carbon. 2015;**95**:460-484. DOI: 10.1016/j.carbon.2015.08.05

[3] Moniruzzaman M, Winey K. Polymer nanocomposites containing carbon nanotubes. Macromolecules. 2006;**39**(16):5194-5205

[4] Paul D, Robeson L. Polymer nanotechnology: Nanocomposites. Polymer. 2008;**49**(15):3187-3204

[5] Sherif A, Qingshi M, Liqun Z, Izzuddin Z, et al. Elastomeric composites based on carbon nanomaterials. Nanotechnology. 2015; **26**(11):112001-112012

[6] Azadmanjiri J, Srivastava V, Kumar P, et al. Two- and three-dimensional graphene-based hybrid composites for advanced energy storage and conversion devices. Journal of Materials Chemistry A. 2018;**6**:702-734. DOI: 10.1039/c7ta08748a

[7] Singh V, Joung D, Zhai L, Das S, et al. Graphene based materials: Past, present and future. Progress in Materials Science. 2011;**56**(8):1178-1271

[8] Stankovich S, Dikin D, Dommett G, Kohlhaas K, et al. Graphene-based composite materials. Nature. 2006; **442**(7100):282-286

[9] Potts J, Dreyer D, Bielawski C, Ruoff R. Graphene-based polymer nanocomposites. Polymer. 2011;**52**(1): 5-25

[10] Kim H, Abdala A, Macosko C. Graphene/polymer nanocomposites. Macromolecules. 2010;**43**(16): 6515-6530

[11] Terrones M, Martín O, Gonzalez M, et al. Interphases in graphene polymer-based nanocomposites: Achievements and challenges. Advanced Materials. 2011;**23**(44):5302-5310

[12] Cai D, Song M. Recent advance in functionalized graphene/polymer nanocomposites. Journal of Materials Chemistry. 2010;**20**(37):7906-7915

[13] Kuilla T, Bhadra S, Yao D, Kim N, et al. Recent advances in graphene based polymer composites. Progress in Polymer Science. 2010;**35**(11):1350-1375

[14] Young R, Kinloch I, Nicolais L. Graphene Composites. Wiley Encyclopedia of Composites. New York City, United States: John Wiley & Sons, Inc; 2011

[15] Sadasivuni K, Ponnamma D, Thomas S, Grohens Y. Evolution from graphite to graphene elastomer composites. Progress in Polymer Science. 2014;**39**(4):749-780

[16] Young R, Kinloch I, Gong L, Novoselov K. The mechanics of graphene nanocomposites: A review. Composites Science and Technology. 2012;**72**(12):1459-1476

[17] Zhu Y, Murali S, Cai W, et al. Graphene and graphene oxide: Synthesis, properties, and applications. Advanced Materials. 2010;**22**:3906-3924

[18] Sun Y, Wu Q, Shi G. Graphene based new energy materials. Energy & Environmental Science. 2011;**4**: 1113-1132

[19] Edwards R, Coleman K. Graphene synthesis: Relationship to applications. Nanoscale. 2013;**5**(1):38-51

[20] Hanssona J, Nilssona T, Ye L, Liu J. Novel nanostructured thermal interface materials: A review. International Materials Reviews. 2017;**63**:22-45. DOI: 10.1080/09506608.2017.1301014

[21] Zhang P, Lv F. A review of the recent advances in superhydrophobic surfaces and the emerging energy-related applications. Energy. 2015;**82**: 1068-1087. DOI: 10.1016/j.energy.2015.01.061

[22] Wasalathilake K, Ayoko G, Cheng Y. Recent Advances in Graphene Research. Chapter 9. Porous Graphene Materials for Energy Storage and Conversion Applications. Rijeka: InTechOpen; 2016. DOI: 10.5772/63554

[23] Divya P, Prithiba A, Rajalakshmi R. Comprehensive review on the nanocomposites of graphene and its derivatives in the application of supercapacitors. Rasayan Journal of Chemistry. 2019;**12**(1):214-231. DOI: 10.31788/RJC.2019.1215039

[24] Zhang Y, Wang Y, Wang C, Gu Y. In: Silvestre N, editor. Thermal Conductivity of Graphene and Its Polymer Nanocomposites: A Review. Advanced Computational Nanomechanics. 1st ed. New York City: John Wiley & Sons; 2016

[25] Garimella S, Persoons T, Weibel J, Gektin V. Electronics thermal management in information and communications technologies: Challenges and future directions. IEEE Transactions on Components, Packaging and Manufacturing Technology. 2017;7:8. DOI: 10.1109/TCPMT.2016.2603600

[26] Moore A, Shi L. Emerging challenges and materials for thermal management of electronics. Materials Today. 2014;**17**(4):163-174

[27] Young R, Kinloch I. Graphene and graphene-based nanocomposites. In: Nanoscience, Nanostructures through Chemistry. Vol. 1. The Royal Society of Chemistry; London, UK; 2013. p. 145-179

[28] Lv L, Dai W, Li A, Lin C. Graphene-based thermal interface materials: An application-oriented perspective on architecture design. Polymers. 2018;**10**: 1201. DOI: 10.3390/polym10111201

[29] Shahil K, Balandin A. Thermal properties of graphene and multilayer graphene: Applications in thermal interface materials. Solid State Communications. 2012;**152**(15): 1331-1340

[30] Goli P, Legedza S, Dhar A, et al. Graphene-enhanced hybrid phase change materials for thermal management of Li-ion batteries. Journal of Power Sources. 2014;**248**:37-43

[31] Bar-Cohen A, Matin K, Narumanchi S. Nanothermal interface materials: Technology review and recent results. Journal of Electronic Packaging. 2015;**137**(4):040803-040817

[32] Mohan V, Lau K, Hui D, Bhattacharyya D. Graphene-based materials and their composites: A review on production, applications and product limitations. Composites Part B Engineering. 2018;**142**:200-220. DOI: 10.1016/j.compositesb.2018.01.013

[33] Li A, Zhang C, Zhang Y. Thermal conductivity of graphene-polymer composites: Mechanisms, properties, and applications. Polymers. 2017;**9**(12): 437

[34] Allen M, Tung V, Kaner R. Honeycomb carbon: A review of graphene. Chemical Reviews. 2010; **110**(1):132-145

[35] Nethravathi C, Rajamathi M. Chemically modified graphene sheets produced by the solvothermal reduction of colloidal dispersions of graphite oxide. Carbon. 2008;**46**(14):1994-1998

[36] Stankovich S, Dikin D, Piner RD, et al. Synthesis of graphene-based nanosheets via chemical reduction of exfoliated graphite oxide. Carbon. 2007; **45**(7):1558-1565

[37] Wang M, Duan X, Xu Y. Functional three-dimensional graphene/polymer composites. ACS Nano. 2016;**10**: 7231-7247. DOI: 10.1021/acsnano.6b03349

[38] Qi G, Yang J, Bao R, et al. Hierarchical graphene foam-based phase change materials with enhanced thermal conductivity and shape stability for efficient solar-to-thermal energy conversion and storage. Nano Research. 2017;**10**(3):802-813

[39] Zhang C, Li A, Zhao Y, et al. Thermal, electrical and mechanical properties of graphene foam filled poly (methyl methacrylate) composite prepared by in situ polymerization. Composites Part B Engineering. 2018; **135**:201-206

[40] Fu Y, He Z, Mo D, Lu S. Thermal conductivity enhancement with different fillers for epoxy resin adhesives. Applied Thermal Engineering. 2014;**66**:493-498

[41] Zhong Y, Zhou M, Huang F, et al. Effect of graphene aerogel on thermal behavior of phase change materials for thermal management. Solar Energy Materials & Solar Cells. 2013;**113**: 195-200

[42] Tang B, Hu G, Gao H, Hai L. Application of graphene as filler to improve thermal transport property of epoxy resin for thermal interface materials. International Journal of Heat and Mass Transfer. 2015;**85**:420-429. DOI: 10.1016/j.ijheatmasstransfer.2015.01.14

[43] Fenga W, Qina M, Feng Y. Toward highly thermally conductive all-carbon composites: Structure control. Carbon. 2016;**109**:575-597. DOI: 10.1016/j.carbon.2016.08.059 1

[44] Ji T, Feng Y, Qin M, Feng W. Thermal conducting properties of aligned carbon nanotubes and their polymer composites. Composites: Part A. 2016;**91**:351-369. DOI: 10.1016/j.compositesa.2016.10.009

[45] Zhao Y, Zhang Y, Bai S. High thermal conductivity of flexible polymer composites due to synergistic effect of multilayer graphene flakes and graphene foam. Composites: Part A. 2016;**85**:148-155. DOI: 10.1016/j.compositesa.2016.03.021

[46] Zhao Y, Wu Z, Bai S. Study on thermal properties of graphene foam/graphene sheets filled polymer composites. Composites: Part A. 2015; **72**:200-206. DOI: 10.1016/j.compositesa.2015.02.011

[47] Kargar F, Barani Z, Salgado R, et al. Thermal percolation threshold and thermal properties of composites with high loading of graphene and boron nitride fillers. ACS Applied Materials & Interfaces. 2018;**10**:37555-37565

[48] Kargar F, Barani Z, Balinskiy M, et al. Dual-functional graphene composites for electromagnetic shielding and thermal management. Advanced Electronic Materials. 2019;**5**: 1800558. DOI: 10.1002/aelm.201800558

[49] Ghosh S, Bao W, Nika D, et al. Dimensional crossover of thermal transport in few-layer graphene. Nature Materials. 2010;**9**(7):555-558

[50] Renteria J, Legedza S, Salgado R, et al. Magnetically-functionalized self-aligning graphene fillers for high-efficiency thermal management applications. Materials and Design. 2015;**88**:214-221

[51] Shahil K, Balandin A. Graphene-multilayer graphene nanocomposites as

highly efficient thermal interface materials. Nano Letters. 2012;**12**(2): 861-867

[52] Shtein M, Nadiv R, Buzaglo M, Regev O. Graphene-based hybrid composites for efficient thermal management of electronic devices. ACS Applied Materials & Interfaces. 2015; **7**(42):23725-23730

[53] Huang Y, Hu J, et al. Manipulating orientation of silicon carbide nanowire in polymer composites to achieve high thermal conductivity. Advanced Materials Interfaces. 2017;**4** (17):1700446. DOI: 10.1002/admi.201700446

[54] Balandin A, Ghosh S, Bao W, et al. Superior thermal conductivity of single-layer graphene. Nano Letters. 2008;**8**: 902-907

[55] Balandin A. Thermal properties of graphene and nanostructured carbon materials. Nature Materials. 2011;**10**: 569-581; 11

[56] Nika D, Ghosh S, Pokatilov E, et al. Lattice conductivity of graphene flakes: Comparison with bulk graphite. Applied Physics Letters. 2009;**94**:203103

[57] Ghosh S, Calizo I, Teweldebrhan D, et al. Extremely high thermal conductivity of graphene: Prospects for thermal management applications in nanoelectronic circuits. Applied Physics Letters. 2008;**92**:15911. DOI: 10.1063/1.2907977

[58] Yu A, Ramesh P, Itkis M, et al. Graphite nanoplatelet-epoxy composite thermal interface materials. Journal of Physical Chemistry C. 2007;**111**: 7565-7569

[59] Min C, Yu D, Cao J, et al. A graphite nanoplatelet/epoxy composite with high dielectric constant and high thermal conductivity. Carbon. 2013;**55**:116-125

[60] Hu J, Huang Y, Yao Y, Pan G, et al. A polymer composite with improved thermal conductivity by constructing hierarchically ordered three-dimensional interconnected network of BN. Advanced Materials Interfaces. 2017;**9**(15):13544-13553. DOI: 10.1021/acsami.7b02410

[61] Dai W, Lv L, Lu J, Hou H, et al. A paper-like inorganic thermal interface material composed of hierarchically structured graphene/silicon carbide nanorods. 2019;**13**(2):1547-1554. DOI: 10.1021/acsnano.8b07337

[62] Ramirez S, Chan K, Hernandez R, et al. Thermal and magnetic properties of nanostructured densified ferrimagnetic composites with graphene – graphite fillers. Materials and Design. 2017;**118**: 75-80. http://dx.doi.org/10.1016/j.matdes.2017.01.018

[63] Suh D, Moon C, Kim D, Baik S. Ultrahigh thermal conductivity of interface materials by silver-functionalized carbon nanotube phonon conduits. Advanced Materials. 2016;**28**: 7220-7227. DOI: 10.1002/adma.201600642

[64] Kholmanov I, Kim J, Ou E, et al. Continuous carbon nanotube–ultrathin graphite hybrid foams for increased thermal conductivity and suppressed subcooling in composite phase change materials. ACS Nano. 2015;**9**: 11699-11707

[65] Levy I, Wormser E, Varenik M, et al. Graphene–graphite hybrid epoxy composites with controllable workability for thermal management. Beilstein Journal of Nanotechnology. 2019;**10**:95-104. DOI: 10.3762/bjnano.10.9

[66] Dmitriev AA, Dmitriev AS, Makarov P, Mikhailova I. New nanocomposite surfaces and thermal interface materials based on mesoscopic

microspheres, polymers and graphene flakes. AIP Conference Proceedings. 1957;**020003**(2018). DOI: 10.1063/1.5034322

[67] Alekseev S, Dmitriev AS, Dmitriev AA, Makarov P, Mikhailova I. Functional energy nanocomposites surfaces based on mesoscopic microspheres, polymers and graphene flakes. IOP Conference Series: Journal of Physics: Conference Series. 2017;**891**: 012361. DOI: 10.1088/1742-6596/891/1/012361

[68] Babenko D, Dmitriev A, Mikhailov V, Mikhailova I. Physical features of Leidenfrost effect on the surface of a grapheme nanocomposite for the problems of thermal and nuclear energy. IOP Conference Series: Materials Science and Engineering. 2018;**447**:012044. DOI: 10.1088/1757-899X/447/1/

[69] Dmitriev A, Valeev R. Graphene nanocomposites as thermal interface materials for cooling energy devices. IOP Conference Series: Journal of Physics: Conference Series. 2017;**891**: 012359. DOI: 10.1088/1742-6596/891/1/012359